Visual Dictionary
of
Firefighters' Tools and Resources

PHILLIP L. QUEEN

THOMSON

DELMAR LEARNING

Australia Canada Mexico Singapore Spain United Kingdom United States

THOMSON

DELMAR LEARNING

Visual Dictionary of Firefighters Tools and Resources

Phillip L. Queen

Vice President, Technology and Trades ABU:
David Garza

Director of Learning Solutions:
Sandy Clark

Acquisitions Editor:
Alison Weintraub

Product Manager:
Jennifer A. Thompson

Marketing Director:
Deborah S. Yarnell

Channel Manager:
Erin Coffin

Marketing Coordinator:
Penelope Crosby

Production Director:
Patty Stephan

Content Project Manager:
Barbara L. Diaz

Art/Design Coordinator:
Nicole Stagg

Technology Project Manager:
Kevin Smith

Technology Project Specialist:
Linda Verde

Editorial Assistant:
Maria Conto

Library of Congress Cataloging-in-Publication Data
Queen, Phillip L.
  Visual dictionary of firefighters' tools and resources / Phillip L. Queen.
    p. cm.
  Includes bibliographical references and index.
    ISBN 1-4018-9790-8 (alk. paper)
1. Fire extinction—Dictionaries.
2. Fire extinction—Equipment and supplies—Dictionaries.   3. Picture dictionaries.   I. Title.
  TH9116.Q4258 2006
  628.9'2503—dc22
                              2006011635

NOTICE TO THE READER

Dedicated to
Dad
the late
Thomas James Queen

"The Original Tool Man"

Contents

Preface

For many years, while writing lessons plans, doing research for a book, or teaching a class, I would always have the frustration of digging through several different resources to find a certain tool or piece of equipment. Sometimes, when I would find the definition of a tool like a "gizmo" or a resource like a "type one engine," I still would not be able to visualize what it looked like.

I wrote this book so that firefighters, instructors, company officers, and chief officers would have a one-stop reference for definitions of various tools and resources, along with photos and diagrams.

Firefighters have many different definitions for the same tool or resource depending on what part of the country they work in or in what context it is being used. I have tried to cross-reference as many definitions as possible.

FEATURES OF THIS BOOK
There are many features incorporated into these pages which will enable you to take full advantage of the content contained within the book:

- **Multi-Purpose Use** This book may be used either a basic firefighting or firefighting tools course for prospective firefighters, or as a handy on the job reference manual.
- **Logically Organized** As with a typical dictionary, this book is organized alphabetically for ease of use. Terms are crossed referenced by names (ex. Litter Basket), as well as by categories

(ex. Baskets, Litter) to help eliminate time consuming searches for definitions.

- **Extensive photos and graphics** are included, as applicable, to assist the learner and provide a visual for the experienced firefighter
- **Resources** are also included, in addition to firefighting tools, to keep firefighters up to date with the various types of assistance available to them and their departments
- **Current technology and resources** are explained and defined, to enable users to keep pace with the latest tools and techniques in the fire service

SUPPLEMENT TO THIS BOOK
For instructors who wish to use this book as a resource in a basic firefighting or firefighting tools courses, an *e.resource CD-ROM* is available, containing the following:

- **Sample Curriculum** illustrates how to organize a tools course, or incorporate into a basic firefighting course
- **Lesson Plans** which outline various topics by subject matter, and enable instructors to use this book as a resource to train incoming firefighters
- **PowerPoint** to correspond to the Lesson Plans
- **Quizzes** correlate to lessons and evaluate student knowledge of the various tools and resources

Order#: 1-4018-9791-6

About the Author

Phillip L. Queen started his career in the fire service in 1968. He became an instructor in 1975 when he received his lifetime teaching credentials fire service related. He has taught firefighters as a consultant and college instructor throughout the country and has written several articles in trade journals on firefighting, command, and management. He has also written a textbook called *Fighting Fires in the Wildland/ Urban Interface* as well as an interactive CD called *I-Zone Basics*.

Queen obtained a bachelor's degree in vocational education and a fire officer certification in the state of California. He now has his own consulting firm that specializes in fire officer training and management. He is an instructor for the Sonoma County Department of Emergency Service, which provides fire protection for Sonoma County, California. Queen also speaks at events and conferences on firefighter survival, fire command, and management training.

Acknowledgments

The fire service is a team-oriented business. Without the team concept we could not function very effectively. While writing this book I have recruited the assistance of many of my teammates. Without the support of my fellow firefighters and co-workers, this project would not have been possible. For the last 37 years I have been fortunate to see the many sides of the fire service and have had the opportunity to work with the best firefighters and instructors in the world.

I would like to gratefully acknowledge the assistance and support of the persons listed below. These individuals shared in the knowledge, reference materials, and support and without them I could not have completed this book.

Mike Paulette (Retired)
California Department of Forestry and Fire Protection
San Diego, California

Dennis Childress
Orange County Fire Authority
City of Orange, California

Vince Hobbs
Hayward Fire Department
City of Hayward, California

Doug McKelvey
Fremont Fire Department
City of Fremont, California

Ted Corporandy
San Francisco Fire Department
City of San Francisco, California

Terry Hein
Tracy Fire Department
City of Tracy, California

In addition, I wish to extend my thanks to the reviewers who shared their knowledge and insightful recommendations over the course of development of this book:

Dennis Childress
Orange County Fire Authority
City of Orange, CA

Benito D. Ramirez
Mount San Antonio College
Walnut, CA

Andrea Walter
Metropolitan Washington Airports Authority
Washington D.C.

I would like to also acknowledge the unnamed persons who generated or shared the information that I have in my files from many years in the fire service.

Without the support of my wife Susan, my family, and friends, this book would never have been accomplished.

A-Frame Hoist   A set of ladders or poles tied off at the tip and at the base into an A-frame configuration that can be used as an emergency hoist point if a pulley and rope are attached to the apex of the ladder or pole triangle (Figure A-1).

Figure A-2A   Absorbent pads will pick up oil from the water.

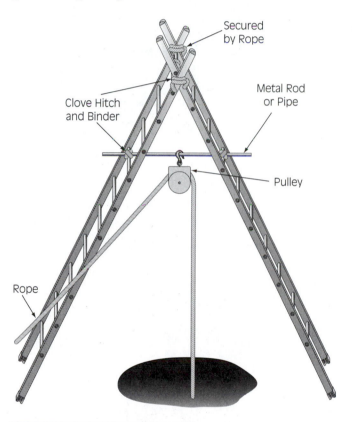

Figure A-1   A-Frame hoist.

A-Frame Ladder   See *Ladder*

Absorbent   Absorbents and absorbent pads (Figure A-2A) soak up any hydrocarbon off the surface of water. Absorbents are specially formulated and licensed for safe and effective use on waterways, holding ponds, contaminated barrels, and sumps. Spilled material is picked up by the absorbent material (Figure A-2B), which acts like a sponge (Figure A-2C). Common absorbent material includes ground-up newspapers, clay, kitty litter, sawdust, and charcoal.

Figure A-2B   Absorbent.

A

Figure A-2C Firefighter putting down absorbent.

Accountability System A system that keeps track of firefighters at an incident. There are several systems used in the fire service.

- **Passport** This is a crew-card system that is tracked on a status board by a monitor (Figure A-3).

Figure A-3 Accountability system in use by firefighters.

- **Tag** In this system, individual firefighters report to staging and gives an identification tag to the staging area manager.

- **Company Officer** This is perhaps the oldest and most commonly used system. Here, the company officer, team leader, or other supervisor is responsible for keeping track of the crew.

Accountability Tags Identification tags to keep track of firefighters through an accountability system (Figure A-4).

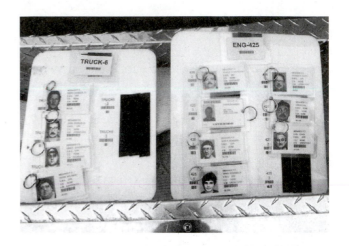

Figure A-4 Accountability tags.

Action Plan Any tactical plan developed by any element of the Incident Command System (ICS) in support of the incident action plan (Figure A-5).

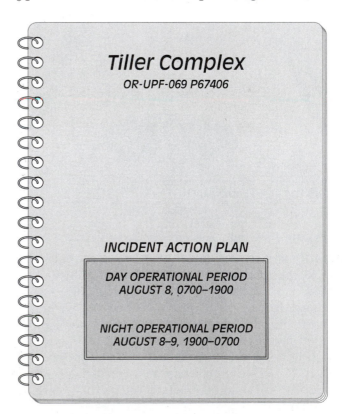

Figure A-5 Action plan.

Adapter A device to make a connection when hose threads do not match or when they are of different sizes (Figure A-6). See also *Hose Appliances*

Figure A-6 Assorted adapters.

Adze Used by firefighters for digging and trenching a line around a wildland fire. The adze consists of a cutting edge fixed at right angles to a handle (Figure A-7A). The bit is curved back toward the handle to facilitate control of the tool during use. The cutting edge is beveled on the side adjacent to the handle. The adze is an ancient tool form and was used in America prior to metal implements. The most common types of adze heads are the flat head (Figure A-7B) and the square head or half flat head

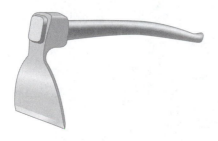

Figure A-7A Adze.

Figure A-7B Flat-head adze, also known as a full-head adze.

(Figure A-7C). Several tools used in the fire service have an adze end.

Figure A-7C Square head or half flat head adze.

Aerial Device An aerial ladder, elevating platform, aerial ladder platform, or water tower designed to position personnel, handle materials, provide egress, and discharge water at elevated locations (Figure A-8).

Figure A-8 Aerial device.

Aerial Fire Apparatus Fire apparatus using mounted ladders and other devices for reaching areas beyond the length of ground ladders (Figure A-9).

Aerial Ignition Device (AID) Used by helicopters for setting backfires, burnouts, or prescribed burns. One such device is called a helitorch (Figure A-10A), an aerial ignition device hung from or mounted on a helicopter to disperse ignited lumps of gelled gasoline. Another aerial ignition device is the delayed aerial ignition device (DAID) (Figure A-10B), which

A

Figure A-9   Aerial fire apparatus.

Figure A-10A   The helitorch is an aerial ignition device (AID).

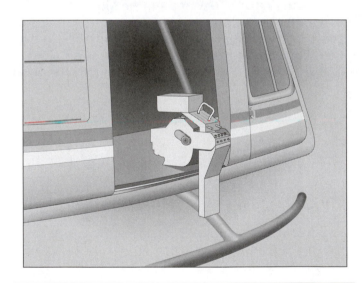

Figure A-10B   Delayed aerial ignition device (DAID) or ping-pong ball system.

disperses polystyrene balls 1.25 inches in diameter containing potassium permanganate. The balls are fed into a dispenser, generally mounted in a helicopter, where they are injected with a water-glycol solution and then dropped through a chute leading out of the helicopter. The chemicals react thermally and ignite in 25 to 30 seconds. It is also called the ping-pong ball system.

Aerial Ladder   A mechanically operated ladder on a turntable attached to a ladder truck chassis and manufactured in various lengths. Because of their heavy-duty use and reach, most of these ladders are very heavy and are constructed of some combination of steel, aluminum, and other metallic alloys in a truss-style configuration for maximum strength. Figure A-11 shows the aerial ladder raising mechanisms.

Aerial Ladder Platform A type of elevating platform that includes the continuous escape capabilities of an aerial ladder (Figure A-12).

Aerial Snorkel Platform The aerial snorkel platform truck allows for the stability and maneuverability that the ladder truck does not provide (Figure A-13).

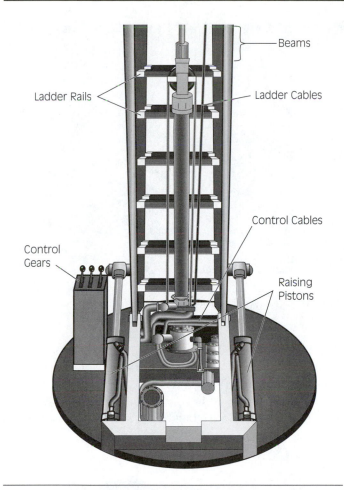

Figure A-11 Aerial ladder raising mechanisms.

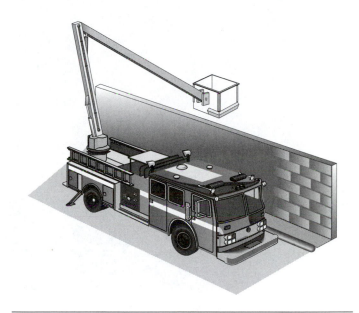

Figure A-13 Aerial snorkel platform.

Figure A-12 Aerial ladder platform.

Aerial Truck Also known as a ladder truck or just a truck. A hydraulically powered ladder mounted on a vehicle that also carries several different length extension ladders, extrication gear, ventilation equipment, and lighting. It may or may not have a bucket or platform on the end (Figure A-14).

Figure A-14 Aerial truck. (Courtesy of John D. Friis.)

Agency Dispatch Center A communications center that has primary jurisdictional responsibility for an incident (Figure A-15).

Figure A-15 Agency dispatch center.

Agent Detection Device Detects the presence of military chemical warfare agents. There are many different types of agent detection devices, such as the one shown in Figure A-16.

Figure A-16 Agent detection device.

Air Bag An inflatable bag, often made of synthetic rubber, used to lift or stabilize heavy objects, such as the high-pressure (Figure A-17A) and the low-pressure air bag (Figure A-17B). Figure A-17C shows a low/medium-pressure bag being used to lift a vehicle.

Figure A-17A High-pressure air bag set.

Figure A-17B Low-pressure air bag set.

Figure A-17C Low/medium-pressure air bag used to lift a vehicle. (Courtesy of Rick Michaelo.)

Air Bottle A thick-walled steel, fiberglass-wrapped aluminum cylinder or composite metal cylinder equipped with a control valve, pressure gauge, and nipple outlet for a high-pressure hose connection. In the fire service these tanks (Figure A-18) contain compressed air up to a pressure of 4,500 pounds per square inch. See also *SCBA*

Figure A-18 Air bottles for breathing apparatus.

Air Chisel An air tool that has been designed specifically for extrication and that is operated off air tanks, compressors, nitrogen bottles, cascade systems, or air brake outlets (Figure A-19). Also known as an air hammer.

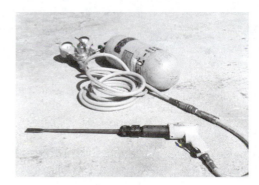

Figure A-19 A high-pressure air chisel kit also known as an air hammer. (Courtesy of Rick Michalo.)

Air Compressor See *Supply Unit*
Air Cylinder See *Air Bottle*
Air Mask Also called a face piece, it is the part of the self-contained breathing apparatus (SCBA) that provides air to the firefighter as well as protection from a hazardous environment. There are two types of air masks—those with a hose attached (Figure A-20A) and without a hose attached (Figure A-20B). There are also specially made lens mounts for air masks that hold prescription lenses. See also *Self-Contained Breathing Apparatus and Airline Mask.*

Figure A-20A Air mask with hose connection.

Figure A-20B Air mask for a self-contained breathing apparatus (SCBA) without a hose connection.

A

Figure A-21 Air monitoring device.

Air Monitoring Device Used to determine oxygen, explosive, or toxic levels of gases in the air (Figure A-21).

Air Packs See *Self-Contained Breathing Apparatus*

Air Purifying Respirators (APR) A respiratory protection that filters contaminants out of the air using filter cartridges. It requires the atmosphere to have sufficient oxygen, in addition to other regulatory requirements. There is the powered respirator (PAPR) (Figure A-22A) and the air purifying non-powered respirator (APR) (Figure A-22B).

Figure A-22A Powered air purifying respirator (PAPR).

Figure A-22B Air purifying respirator (APR).

Air Supply Unit A machine to refill exhausted compressed air bottles (Figure A-23).

Figure A-23 Air supply unit (compressor/purifier).

Air Tank See *Air Bottle*

Aircraft There are a variety used in the fire service today. Aircraft are either fixed-wing or rotary.

Fixed-Wing Aircraft Used for dropping fire retardant (Figure A-24) or water onto wildland

Figure A-24 Aircraft making a retardant drop.

fires, to reconnaissance or as a lead plane to direct air traffic flying over a fire.

Air Tankers Any fixed-wing aircraft capable of transporting and delivery of fire suppressant or retardant materials. There are four categories of air tankers:

- **Type-One Air Tanker** The largest air tanker (Figure A-25A) carries up to 3,000 gallons of fire retardant. Aircraft such as the Lockheed C-130 four-engine turboprop can cruise at 275 mph. The C-130 has two applications— conventional and modular airborne firefighting system (MAFFS), a system that slides in the back where two large pipes hang behind and below the open cargo door. Pressurized air forces retardant out of the pipes. Once a drop is started, the entire load is released. Other type-one air tankers include the P-3 and DC-7 aircraft.

Figure A-25A Type-One air tanker.

- **Type-Two Air Tanker** Shown in Figure A-25B, it has a minimum capacity of 1,800 gallons of retardant. The P2-U, SP2-H, and some DC-7s are considered type-two air tankers. There is also the water scooping aircraft (Figure A-25C).

Figure A-25B Type-Two air tanker.

Figure A-25C Type-Two water scooping airtanker.

- **Type-Three Air Tanker** Shown in Figure A-25D, it has a minimum capacity of 800 gallons of retardant. The S-2 is used by the California Department of Forestry and Fire Protection. The CL-215t and the CL-215 are also type-three air tankers.
- **Type-Four Air Tanker** Shown in Figure A-25E, it is the smallest and must carry a minimum of 100 gallons of fire retardant. Aircraft such as the Thrush and the Dromader are type-four air tankers.

A

Figure A-25D Type-Three air tanker.

Figure A-25E Type-Four air tanker.

- **Lead Plane, Air Tactical, and Bird Dog** Shown in Figure A-26, these are small aircraft such as the Baron, the ov10 Bronco, 02a, Aero

Figure A-26 Air attack or air tactical plane.

Commander, and the Aerostar. They carry the air tactical group supervisor (ATGS), the person responsible for directing and coordinating airborne aircraft operations and managing air space for an incident. The lead plane is an aircraft used to make trail runs to check wind, smoke conditions, and topography to lead air tankers to targets and supervise their drops.

Rotary Wing Aircraft (Helicopters) Used for rescue, medical evacuation, and dropping water or fire retardant onto wildland fires. Helicopters (Figure A-27) are also used to transport crews to and from the fire lines, where some are dropped by rope near the fire line or rescue area. Helicopters are also used for reconnaissance and directing air and ground crews at a fire or rescue situation.

Figure A-27 Helicopter.

- **Helitanker** A specialized use of crane helicopters. These are very large, twin-engine, single-rotor aircraft. The helitanker configuration (Figure A-28A) carries a 2,600-gallon aluminum tank that delivers 1,800–2,600 gallons of water with pinpoint accuracy onto fires. It also has the capability to pick up water from a portable tank (Figure A-28B). Minimum standards for a helitanker are:
    Fixed tank
    Air tanker board certified
    1,100-gal. min. capacity
- **Type-One Helicopter (Heavy)** Carries approximately 1,000 gallons of water or retardant in a bucket suspended approximately 50–100 feet below the helicopter (Figure A-28C). Minimum standard, for a type-one helicopter are:
    Seats   16
    Card weight capacity (lbs.)   5,000
    Minimum retardant (gal.)   700

Figure A-28A Helitanker.

Figure A-28B Helitanker picking up water from a portable tank.

Figure A-28C Type-One helicopter.

- **Type-Two Helicopter (Medium)** Carries approximately 400 gallons of water or retardant in a bucket suspended approximately 50–100 feet below the helicopter (Figure A-28D). Minimum standards for a type-two helicopter are:

Seats	10
Card weight capacity (lbs.)	2,500
Minimum retardant (gal.)	300

Figure A-28D Type-Two helicopter.

- **Type-Three Helicopter (Light)** Carries approximately 90 gallons of water or retardant in a bucket suspended approximately 50–100 feet below the helicopter (Figure A-28E). Minimum standards for a type-three helicopter are:

Seats	5
Card weight (lbs.)	1,200
Minimum retardant (gal.)	100

Figure A-28E Type-Three helicopter.

A

- **Type-Four Helicopter** Very maneuverable in confined areas and has great visibility for long line work (Figure A-28F). These helicopters are very economical to operate. Minimum standards for a type-four helicopter are:

Seats	3
Card weight (lbs.)	600
Minimum retardant (gal.)	75

Figure A-28H   Coast Guard rescue helicopter.

Figure A-28F   Type-Four helicopter.

- **Rescue Helicopters** There are a variety of rescue helicopter and medical evacuation helicopters that are civilian or government-owned (Figure A-28G). The fire service also uses helicopters from other government agencies such as the military and Coast Guard (Figure A-28H).

Aircraft Rescue and Firefighting Designation of firefighting personnel and equipment specializing in airport/aircraft emergencies based at or near an airport (Figure A-29).

Figure A-29   Aircraft rescue firefighting vehicle (ARFF).

Airline Mask Worn by firefighters and connected to an air supply outside a contaminated area (Figures A-30A and A-30B).

Airport Crash Truck The crash and rescue equipment includes: water; foam (light water); and halon 1211 foam, made with a chemical designed to smother fire (Figure A-31).

Alidade or Alidade Tables A sighting device used by lookouts to determine the horizontal bearing and sometimes the vertical angle of a fire from a lookout (Figure A-32). Lookout towers provide fixed-point fire detection from different fire tower locations, using alidade tables, binoculars, and topographic maps to triangulate the precise location of a fire.

All-Purpose Hook   See *Hooks*

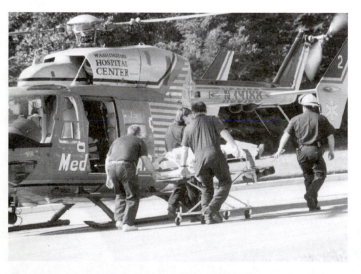

Figure A-28G   Helicopter medi-vac. (Photo courtesy of Fred Schall, Sterling Park Rescue Squad.)

Figure A-30A Airline mask to an outside air supply.

Figure A-30B Airline mask with an emergency escape cylinder.

Figure A-31 Airport crash truck. (Courtesy of Symtron Systems, Inc.)

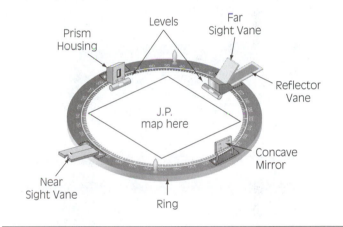

Figure A-32 Alidade.

Anchor Plate Rigging plates that keep ropes and equipment from becoming entangled (Figure A-33).

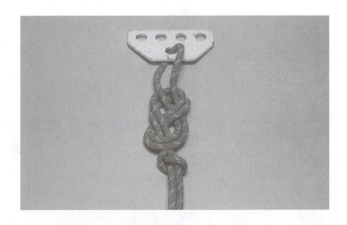

Figure A-33 Anchor plate.

Anchor Pulley A single-directional pulley for rope rescue operations (Figure A-34).

A

Figure A-34 Anchor pulley.

Anchor Sling Synthetic-fiber flat or tubular webbing configurations (Figure A-35A) that wrap around anchors such as trees (Figure A-35B) for the purpose of attaching system components.

Figure A-35A Anchor slings—synthetic anchor straps.

Figure A-35B Anchor sling around a tree.

Anemometer General name for instruments designed to measure wind speed. The handheld anemometer that comes in the belt weather kit is also called a wind speed indicator (Figure A-36).

Figure A-36 Anemometer—handheld wind speed indicator that comes in the belt weather kit.

Annunciator Panel A visual and audible indication of system fire alarms for a building, provided for the rapid identification of the location of an alarm initiation. The fire service relies on alarm annunciation to determine fire origin in large, multistoried, or complex structure. Also known as the System Control Panel (Figure A-37).

Figure A-37 Annunciator panel (AP).

When checking the annunciator panel the firefighter should check:

1. That all indicators on the panel are in a "normal" condition and not in "trouble" mode.
2. That the panel is receiving AC power.
3. That all detection devices are present and properly mounted, not hanging from the ceiling.
4. That records are provided indicating that the system has been tested periodically.

Anti-Torque Device (ATD) Another name for swivels and snap links. This device is used for untangling twisted ropes and equipment (Figure A-38).

Figure A-38 Anti-torque device (ATD).

Articulating Boom Ladder An apparatus with a series of booms and a platform on the end, maneuvered into position by adjusting the various boom sections to position the platform at the desired location (Figure A-39).

Figure A-39 Articulating boom.

Ascenders Used by rescue workers as mechanical prusiks in systems and rappels (Figure A-40).

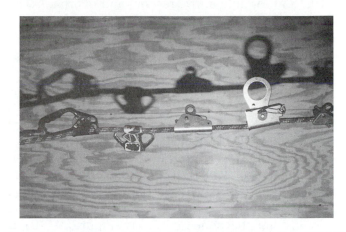

Figure A-40 Ascenders, also called rope grabs.

A-Tool Beveled, triangular lock puller found in many fire department service tools (Figure A-41).

Figure A-41 The A-Tool is not a separate tool, but is usually attached to the end of a bar or forcible entry tool.

Attack Line A line of hose preconnected to the pump of a fire apparatus and ready for immediate use in attacking a fire. In contrast, supply lines connect a water supply with a pump or to feeder lines extended from a pump to various points around the perimeter of a fire (Figure A-42).

Figure A-42 Attack line.

A

Axes Firefighters use several types of axes for various firefighting and rescue operations, such as for ventilation, forcible entry, rescue, and overhaul. The following is a list of axes used in the fire service today.

- **Brush Hook** Designed for fire line clearing in heavy brush areas. It is constructed from carbon tool steel with a No. 1 hickory handle. The overall length of the head is 8 7/8 inches with a cutting edge 7 5/8 inches long. The handle is 36 inches, the overall length is 42 inches, and the weight is 5 pound, The most common brush hooks used by the fire service are the strap hook and the blood brush hook (shown in Figure A-43A). The brush hook is a dangerous tool; if not sharpened correctly (Figure A-43B), it can cause injuries.

- **Crash Axe** One-piece drop-forged steel axe with a steel handle (Figure A-43C), rubber insulated to 20,000 volts. It is lightweight, heavy-duty, and well balanced. Designed to go through heavy metals and for use when an emergency requires prompt results. The approximate weight is 2.5 pounds, while the overall length is 152 inches.

- **Double-Bit Axe**: A two-handed axe with two cutting bits.The most common double-bit axes are the Adirondack pattern and the California Redwood pattern (Figure A-43D). The bits are

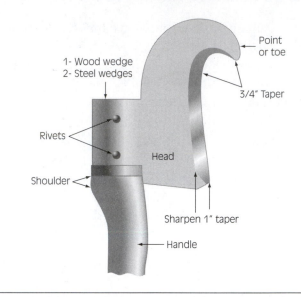

Figure A-43B Brush hook sharpening procedure.

Figure A-43C Crash axe.

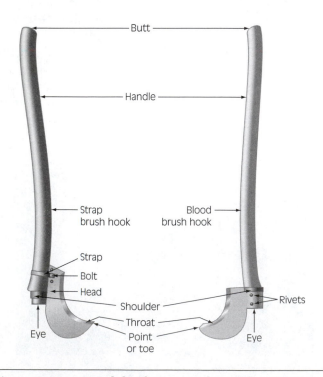

Figure A-43A Brush hook types and part identification.

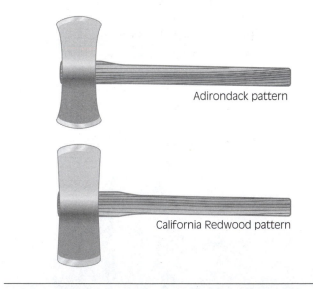

Figure A-43D Common patterns of the double-bit axe.

usually identical. The weight is 3 to 6 pounds. This axe needs to be sharpened correctly to be effective (Figure A-43E) and the head needs to be mounted to the handle at a 90-degree angle with

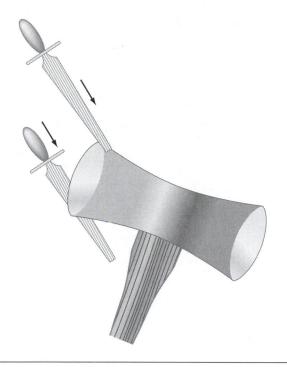

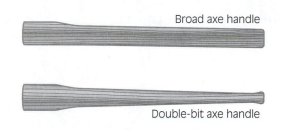

Figure A-43G   Axe handles.

cutting, and pulling up boards. It generally weighs from 3.5 to 8 pounds and has a 4.5- to 6-inch cutting edge (Figure A-43H). There are three basic patterns for the fire axe—Hunts Underhill, and Lippincott (Figure A-43I).

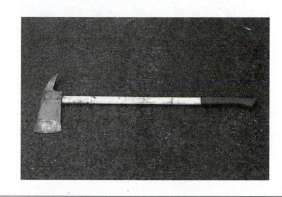

Figure A-43E   Double-bit axe sharpening angle.

wooden, metal, or plastic wedges (Figure A-43F). There are two types of handles for double-bit axes: one is the broad axe handle and the other is the double-bit axe handle (Figure A-43G).

- **Fire Axe or Pick-Head Axe**   A fire axe has a pick head and a blade. The fire axe is used for general prying and chopping, roof ventilation,

Figure A-43H   Pick-head or fire axe.

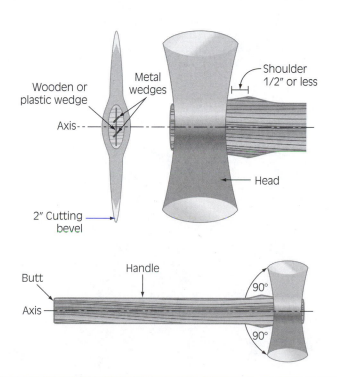

Figure A-43F   Double-bit axe parts.

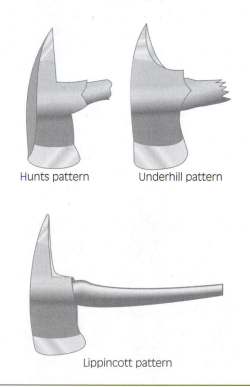

Figure A-43I   Common fire axe patterns.

A

- **Main Axe** The main axe is like the Pulaski (see below) but lighter and more streamlined (Figure A-43J).

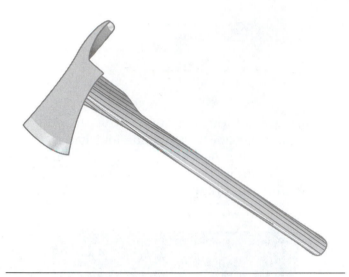

Figure A-43J Main axe.

- **Maul Axe** A heavy axe used for driving wedges and breaking material. The maul head is broad and heavy, made to power apart wood with their wide wedge shape. Its weight is ideal for breaking up and chopping. The patterns used in the fire service most often are called rafting and maul patterns (Figure A-43K).

Maul   Rafting

Figure A-43K Maul axe patterns.

- **Pry Axe** A multi-purpose tool designed for prying, twisting, chopping, metal cutting, and ramming (Figure A-43L). This efficient tool lets firefighters force entry, rescue, and ventilate. See also *Biel Tool*

- **Pulaski or Mattock Axe** A cutting tool with an axe bit on one side and a Mattock bit on the other side (Figure A-43M). The Pulaski is an ideal tool for chopping, grubbing, or trenching. The Pulaski must be sharpened correctly to be used effectively as a firefighting tool (Figure A-43N).

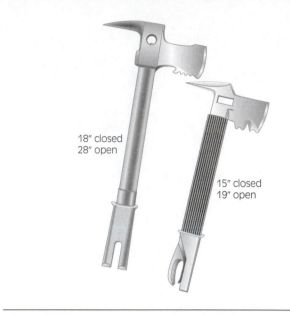

18" closed
28" open

15" closed
19" open

Figure A-43L Pry axe.

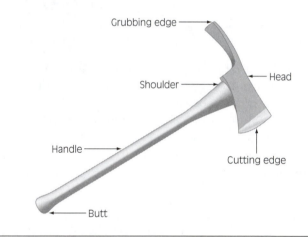

Grubbing edge

Head

Shoulder

Handle

Cutting edge

Butt

Figure A-43M Pulaski axe parts.

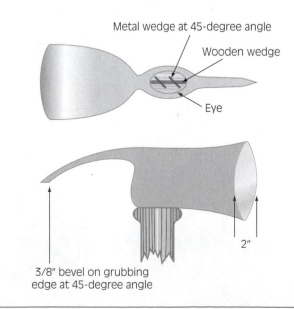

Metal wedge at 45-degree angle

Wooden wedge

Eye

3/8" bevel on grubbing edge at 45-degree angle

2"

Figure A-43N Pulaski axe sharpining procedures.

• **Quick Axe** A combination forcible entry tool that can be used as an axe, spanner wrench, pry tool, and hammer (Figure A-43O).

Figure A-43O Quick axe.

• **Single-Bit Axe or Flat-Head Axe** An axe having only one bit (Figure A-43P) with a cutting edge that is generally beveled on both sides. The standard single-bit axe is a two-handed general purpose tool used for felling, wood cutting, trimming, and most other chopping tasks. The poll opposite the cutting bit serves primarily for the purpose of weight and balance but is sometimes hardened for use as a maul. There are four basic bevel types used in the fire service single-bit axe: the common bevel, phantom bevel, hollow bevel, and no bevel (Figure A-43Q). The flat-head or single-bit axe is used for several tasks in the service. Figure A-43R shows a firefighter breaking a window, while Figure A-43S shows the flat-head axe being used with the Halligan tool. Several fire departments use the combination of the flat-head

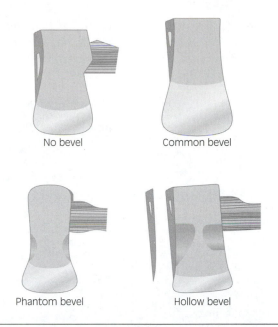

No bevel   Common bevel

Phantom bevel   Hollow bevel

Figure A-43Q Flat-head axe's most common bevels.

Wind direction

Break window with the flat side of the axe

Figure A-43R Flat-head axe being used to break window with its flat side.

Figure A-43P Flat-head or single-bit axe.

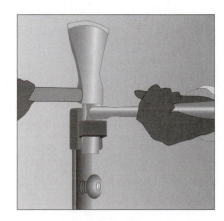

Figure A-43S Flat-head axe in use with a Halligan tool.

A

axe and the Halligan tool called the irons (Figure A-43T). The single-bit or flat-head axe handles come in two types—the fawn foot and the single-bit axe handle (Figure A-43U).

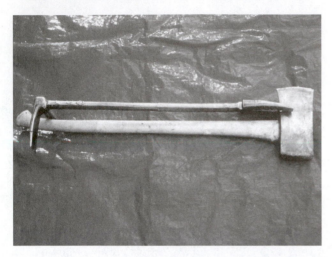

Figure A-43T Flat-head axe with the Halligan tool form the tool set known as the irons.

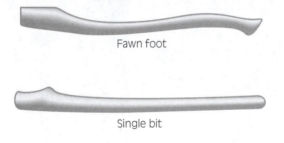

Fawn foot

Single bit

Figure A-43U Handles for single-bit axe.

• **Swedish Brush Axe** The Swedish steel blade cuts brush and saplings. This axe is lighter than regular axes or machetes. The blade can be quickly replaced in the field. Its length is 28 inches and its weight is only 2 pounds (Figure A-43V).

Figure A-43V Swedish brush axe.

• **Trench Axe or Adze Axe** A narrow-bit adze in combination with an axe bit (Figure A-43W). The shank of the adze bit is curved back toward the handle rather than straight as in the Mattock or Pulaski.

Figure A-43W Trench or adze axe.

Axe Sheath A cover for axe heads usually made of leather (Figure A-44).

Figure A-44 Axe sheath.

Azimuth Circle A circle graduated in 360 degrees in a clockwise direction from true (astronomic) north (Figure A-45). Used by firefighters on alidades to pinpoint fires in the wildlands. See also *Alidade*

Figure A-45   Azimuth circle.

Figure A-46   Azzard.

Azzard   Provides connections to towers and other structures when ascending with a lead climbing technique. The design allows quick, one-handed connection to the structure for maximum security when placing belay protection (Figure A-46).

B Plus Suit   See *Protective Clothing*

B Suit   See *Protective Clothing*

Back Fire Torch   A flame-generating device, such as a fount containing diesel oil or kerosene and a wick or a backpack pump serving a flame-jet. A back fire torch is also a flare manufactured for firefighting operations (Figure B-1). See also *Drip Torch*

Figure B-1   A 10-minute backfiring torch.

Backpack Applicator   Used to apply foam, similar to the backpack pump (Figure B-2).

Figure B-2   Backpack applicator.

Backpack Pump   Five-gallon galvanized steel or stainless steel tanks that fit on one's back and use a piston-type pump (Figure B-3A). There are also tanks constructed from high-strength nylon, coated with an extremely tough vinyl (Figure B-3B). See also *Bladder Bag*

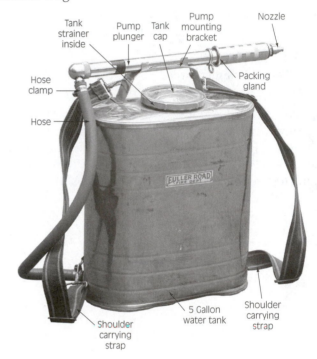

Figure B-3A   Backpack pump.

Figure B-3B   Backpack pump fabric tank or bladder bag.

Bags The fire service uses several different types of bags, usually for storage and equipment protection.

- **Air Bag** See *Air Bag*
- **Hydrant Bag** Holds hydrant tools such a spanner wrenches, hydrant wrenches, and rubber mallets. It attaches to the hydrant end of the hose and does not need to be removed when charging the line (Figure B-4A).

Figure B-4C RIT bag with tools, air bottle, and mask.

Figure B-4A Hydrant bag.

- **Mask Bag** Protects an air mask from the elements (Figure B-4B).

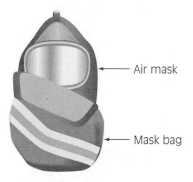

Air mask

Mask bag

Figure B-4B Mask bag.

- **RIT Bag** Carries the tools used by the rapid intervention team (RIT) (Figure B-4C).
- **Rope Bag** Provides protection for the rope and keeps it neatly stored and easily deployable (Figure B-4D).
- **Rope Stuff Bag** Rope is stuffed into this bag that has a hole on the bottom where a knot is tied. This procedure is also called bagging a rope (Figure B-4E).

Figure B-4D Rope bag.

B

Figure B-4E   Rope bag stuff.

Figure B-4F   Salvage bag.

Figure B-4G   Salvage bag or carry-all.

- **Salvage Bag**   Used during salvage and overhaul to carry debris (Figure B-4F). Shown here are firefighters using a salvage bag to remove debris (Figure B-4G).

Balanced Pressure Demand-Type Foam Proportioner System   Uses one or two foam pumping methods in which foam is pumped under pressure into a metering chamber that balances the pressure of the concentrate and water, thereby controlling the flow of the foam solution (Figure B-5).

Bale Hook   A hook-shaped tool for moving bales or other items, also used for hauling stuffed furniture.

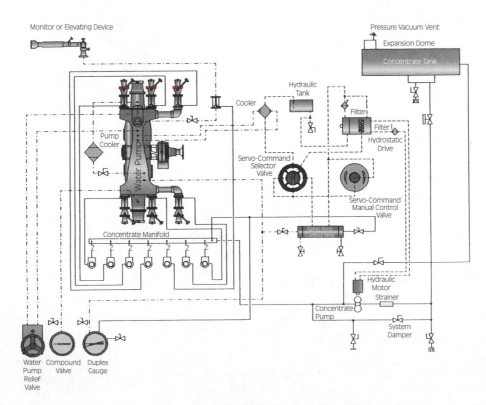

Figure B-5   Balanced pressure demand-type foam proportioner system.

Shown in Figure B-6 are the most common bale hook patterns. See also *Hooks*

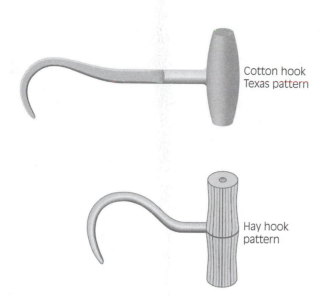
Cotton hook
Texas pattern

Hay hook
pattern

Figure B-6 Bale hook patterns.

Ball Valve See *Valves*

Ballistic Line Gun Uses gunpowder to project a rod or rubber ball carrying a pilot cord to which a rope can be attached (Figure B-7).

Figure B-7 Ballistic line gun.

Bam-Bam Tool An automotive dent puller used by firefighters to pull or open locks (Figure B-8).

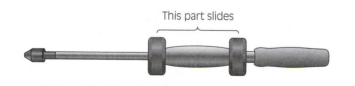
This part slides

Figure B-8 Bam-bam tool.

Bambi Bucket A collapsible bucket slung below a helicopter that dips water from a variety of sources for fire suppression (Figure B-9).

Bangor Ladder See *Ladder*

Bars There are hundreds of different types of bars used in the fire service today. Many of these bars are

Figure B-9 Bambi bucket.

made at the fire stations and modified for a particular use and area. The following are just a few examples:

• **Chicago Patrol Bar** A Halligan-style tool with a hammerhead configuration on the adze end and a pike or pick point (Figure B-10A).

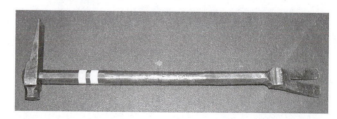

Figure B-10A Chicago patrol bar or Chicago door opener. (Photo courtesy of Ted Corporandy.)

• **Crow Bar** Approximately 48 to 54 inches long and weighing around 12 pounds, used during auto extrication, overhaul, heavy rescue, and ventilation operations (Figure B-10B). Also called a pinch bar or pry bar.

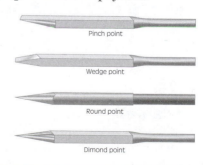
Pinch point

Wedge point

Round point

Dimond point

Figure B-10B Crow bar or pry bar.

B

- **Duck Bill Bar** A forcible entry tool with a tapered head designed to be placed in the shackle of a padlock. When hit with a mallet or the head of an axe, it spreads the shackle open (Figure B-10C).

Figure B-10F Hayward lock breaker, also known as an L.A claw tool or a New York claw tool.

- **Hook and Claw Tool** A forcible entry tool that has a fork on one end and a 5-inch hook or pointed claw on the other (Figure B-10G).

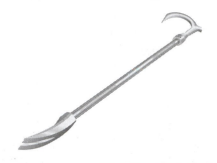

Figure B-10G Hook and claw tool.

- **Hux Bar** A hydrant wrench and forcible entry tool made of chromed tubular steel, with a very thin, flat hydrant tool at one end and a curved, tapered prying point with a small fulcrum at the other (Figure B-10H).

Figure B-10H Hux bar.

- **Kelly Bar** Often used in a wedge and sledge combination with a flat-head axe. The adze end extends into a protruding hammerhead for use as a striking tool or to receive hammer blows. The claw has a recessed nail puller and is parallel-slotted to slip over locks, hasps, and many gas shutoffs (Figure B-10I).

Figure B-10C Duck bill lock breaker forcing a heavy-duty padlock.

- **Fire Bar** Forcible entry tool for opening deadbolts and security bars (Figure B-10D).

Figure B-10D Fire bar.

- **Halligan Tool** From the prying group, it is a 30-inch forged steel tool with three primary parts: the adze end, the pike end, and the fork end (Figure B-10E).

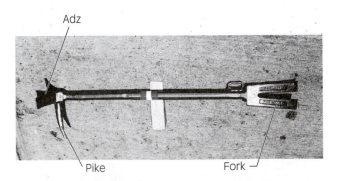

Adz

Pike　　　　　　Fork

Figure B-10E Halligan tool.

- **Hayward Lock Breaker** Consists of three heads used for forcible entry, ventilation, overhaul, and vehicle extrication (Figure B-10F).

Figure B-10I Kelly bar.

- **Officer's Tool** A forcible entry tool made of steel with a block-type head welded to the shaft. The shaft is formed into a pry bar capable of slipping into tight spaces. The head of the tool has a flat driving surface with an A-type lock puller machined into the adze of the blade. This tool also contains a double-ended "key" tool (or Kerry key) and a shove knife. The kit is stored inside the grip of the shaft (Figure B-10J).

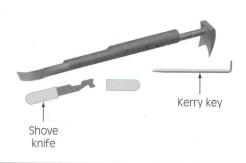

Figure B-10J Officer's tool.

- **Pinch Bar** See *Crow Bar*
- **Pro Bar** Looks like a Halligan tool but has a fork, adze, and point for penetration (Figure B-10K).

Figure B-10K Pro bar.

- **Quic Bar** A three-piece bar whose ends are forged in high-alloy aircraft steel and are tapered to fine points to facilitate entry. It has wedge ends that are pinned through for maximum strength. Available in 24-, 30-, or 42-inch lengths (Figure B-10L).

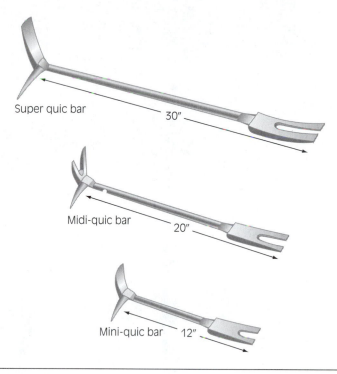

Figure B-10L Quick bar.

- **Ram Bar** A forcible entry tool that uses a ramming power stroke on a fork end (Figure B-10M), like a piston.

Figure B-10M Ram bar.

- **Rescue Bar** A forcible entry and prying tool with a flat chisel end for getting into tight spaces (Figure B-10N).
- **Rex Tool** Similar to a lock-pulling tool, a rex tool is shaped like a U with tapered blades that bite into lock cylinders of different shapes and sizes. The opposite end is like a chisel, used to drive rim locks after their cylinders are removed (Figure B-10O).

B

Figure B-10N   Rescue bar.

Figure B-10O   Rex tool being used to pull a tubular bolt lock.

- **San Francisco Bar**   A forcible entry tool that is 30 inches long with a half round on the back side of the fork to create a pronounced fulcrum (Figure B-10P).
- **Truckman's Bar**   Also called a truckie tool, it is a forcible entry tool that has three striking surfaces. This tool also has an A-type lock puller that can be used as a gas shut-off, edge tool, nail puller, hammer, and pry bar (Figure B-10Q).
- **Vernon Bar**   Used for peeling off roofing tar paper and tile shingles (Figure B-10R).

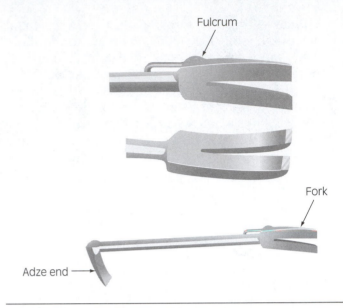

Figure B-10P   San Francisco bar.

Figure B-10Q   Truckman's bar, also called a truckie tool.

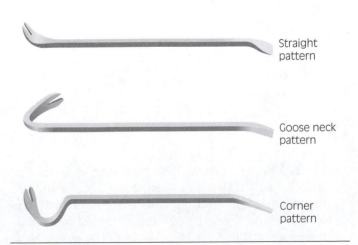

Figure B-10R   Vernon bar.

- **Wrecking Bar**   A small bar used for prying and forcible entry, often mistakenly called a crow bar. Also called the carpenter's prying bar, carpenter's ripping bar, nail bar, or ripping bar (Figure B-10S).

Figure B-10S   Wrecking bar.

B

• **Zak Bar** A forcible entry tool with a hammer-head design, made of lightweight, heat-treated aircraft stainless steel. The 1.25-inch steel tube is designed to bend but not snap under force (Figure B-10T).

Figure B-10T Zak bar.

Bar Spreader A device that is attached to a harness to keep gear from tangling (Figure B-11).

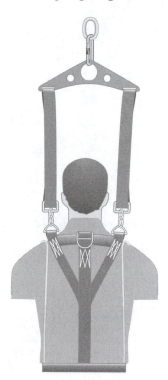

Figure B-11 Bar spreader.

Barracuda Saw See *Saw*

Barricade Tape A ribbon or tape available in different colors to block off a scene or denote a hazard (Figure B-12). See also *Flagging*

Barrel Strainer A barrel-shaped strainer placed over the end of a suction hose to prevent debris from entering the pump (Figure B-13).

Barron Tool A wildland firefighting tool that has a rake and hoe blade set at a 65-degree angle to shear rapidly, used for cutting light roots, matted grass, and small brush without cutting into the mineral soil. Its tines rarely clog and the tool can be used to hold and

Figure B-12 Barricade tape.

Figure B-13 Barrel strainer.

drag burning material when you are setting fires for burnout operations. Also called the California fire tool (Figure B-14).

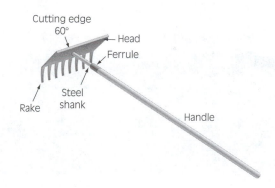

Figure B-14 Barron tool or California fire tool.

Base Radio See *Radio*

Basket Suction Strainer A suction strainer shaped like a basket that fits on the end of a suction hose to prevent debris from entering the pump (Figure B-15).

Figure B-15 Basket suction strainer.

Basket   The area of a snorkel in which firefighters work (Figure B-16).

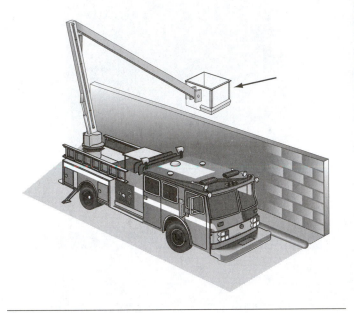

Figure B-16   Basket.

Battering Ram   A heavy metal bar used by police officers and firefighters to break down doors (Figure B-17). Also called a door breaker.

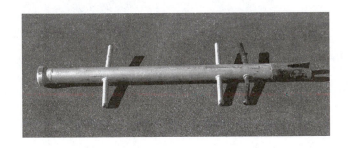

Figure B-17   Battering ram.

Beam   See *Ladder*

Bed Ladder   See *Ladder*

Belt   There are many types of belts and accessories used in the fire service today.

- **Bailout Belt**   A last-chance emergency bailout uniform belt with an attached D-ring (Figure B-18).
- **Belt Cutter**   A V-shaped device with a sharp blade for cutting seat belts or webbing (Figure B-19).
- **Pompier Belt**   Ladder/escape belts (Figure B-20A) used both as a positioning device for a person on a ladder (Figure B-20B) as well as by the wearer as an emergency self-rescue device.

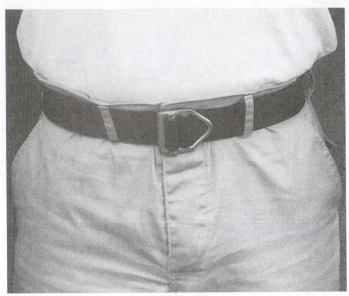

Figure B-18   A last-chance emergency bailout uniform belt.

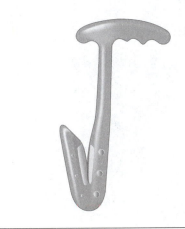

Figure B-19   Belt cutter also called a V-blade or rescue knife.

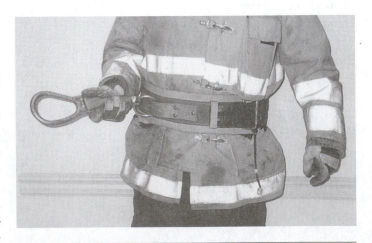

Figure B-20A   Firefighter's rescue or pompier or ladder belt.

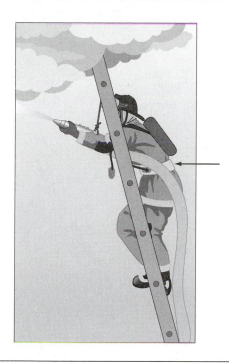

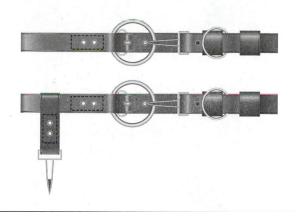

Figure B-21 Truckman's belt with accessories.

Figure B-20B Ladder belt being used to secure firefighter to ladder.

- **Truckman's Belt** Usually made of leather and sometimes webbing to attach a D-ring and snap hook (Figure B-21).

- **Belt Weather Kit** A small kit designed to fit on a belt that includes a compass along with several specialized pieces of equipment that determine the current wind speed and direction, air temperature, and humidity. A notebook and pencil are included for documenting results. None of the equipment is electronic (Figure B-22).

Biel Tool A multi-purpose forcible entry and rescue tool for forcing doors and windows, prying, and twisting hasps and locks (Figure B-23). Also known as pry axe.

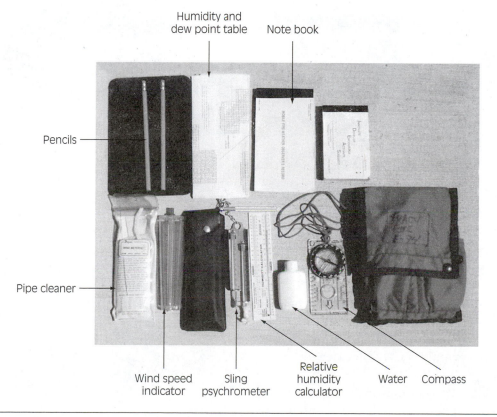

Figure B-22 Belt weather kit.

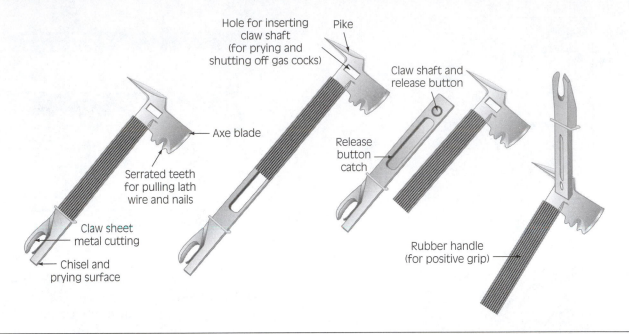

Figure B-23 Biel tool.

Binoculars Linked pair of small telescopes used to look at distant objects through use of a magnifying lens for each eye. Firefighters use binoculars for viewing objects at a distance if it is a safety concern such as at a hazardous materials response (Figure B-24).

Figure B-24 Binoculars. (Photo courtesy of Terry Hein)

Biohazard Container A puncture-resistant polyethylene or thermoplastic container that is used to safely dispose of biohazards (Figure B-25).

Bladder Bag A collapsible backpack portable sprayer made of neoprene or high-strength nylon fabric fitted with a pump. See *Backpack Pump*

Blocks There are several types of blocks used in the fire service today.

- **Shoring Block** A section of 2×4-, 4×4-, or 6×6-inch wood used to crib up vehicles or objects. Shoring blocks can be of different shapes

Figure B-25 Biohazard containers.

and sizes and are made from several types of materials (Figure B-26). Also called cribbing.

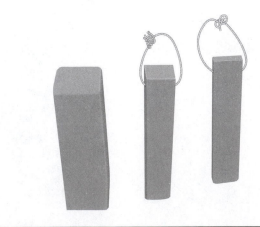

Figure B-26 Shoring blocks, also called cribbing.

B

• **Snatch Block**   A pulley that opens on one side to insert a rope, thereby avoiding the necessity of threading the rope from one end (Figure B-27).

Figure B-27   Snatch block.

Blow Down Fitting   A device that removes water from a booster hose so it will not freeze (Figure B-28).

Figure B-28   Blow down fitting.

Blower   Also called smoke blowers and smoke ejectors (Figure B-29A), used to remove smoke and gases from an area through positive pressure (Figure B-29B).

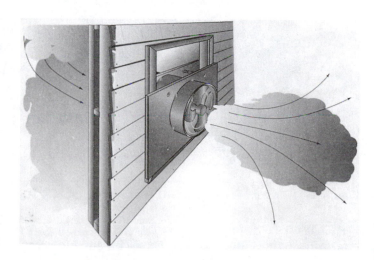

Figure B-29A   Blower or smoke ejector.

Figure B-29B   Blower being used for positive-pressure ventilation.

Blowers can also be used to force fresh air into a confined space (Figure B-29C). There are several types of blowers, such as the gas-powered blower with ventilation hose, the electric-powered blower, and the electric-powered blower with ventilation hose (Figure B-29D).

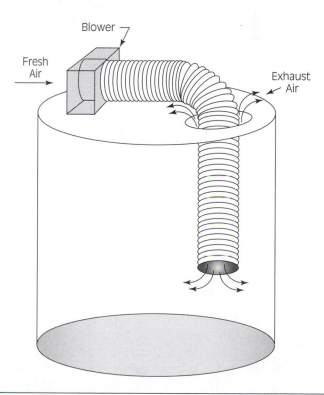

Figure B-29C   Blower use to pump in fresh air and pump out bad air in a confined space.

B

Gas-powered with ventilation hose   Electric-powered   Electric-powered with ventilation hose

Figure B-29D Blowers.

Boats There are several types of boats used in the fire service today, divided into five categories.

1. **Air** A propulsion system using an aviation propeller or a ducted fan to generate thrust from the engine with an on-plane draft of 0 to 12 inches. Typical boats in this category include "swamp boats" and surface boats.

2. **Jet** A propulsion system using a water pump to generate thrust with an on-plane draft of 6 to 12 inches. Jet boats are susceptible to damage from floating debris.

3. **Propeller** A propulsion system using a propeller to generate thrust with an on-plane draft of 18 to 24 inches.

4. **Non-Powered** A non-motorized vessel capable of safely transporting rescuers or victims, such as a raft, skiff, johnboat, and so on.

5. **Powered** A motorized vessel capable of safely transporting rescuers or victims, such as an inflatable rescue boat (IRB), a rigid hull inflatable rescue boat (RIHB), a rigid hull boat, a personal water craft (PWC), airboat, and so on.

   • **Fire Boat** There are numerous types of fire boats (Figure B-30A), from very large to very small. FIRESCOPE, a California Incident Command Equipment standard, describes a type-one fire boat as having a pumping capacity of 5,000 gallons per minute, a type-two fire boat as having a pumping capacity of 1,000 gallons per minute; and a type-three fire boat as having a pumping capacity of 250 gallons per minute.

   • **Ice Boat** A floating platform used for rescue on ice (Figure B-30B).

Figure B-30A Fire boat.

Figure B-30B Rescue ice boat. (Photo courtesy of Halfmoon-Waterford Fire District 1.)

   • **Rescue Boat** There are many types of rescue boats, from the very large used at sea in a large port to very small boats such as rubber ice boats.

Bobbin Descent Control Device Uses friction through the rope passing over a series of non-rolling metal bobbins (Figure B-31). Friction is adjusted using the handle for leverage or by adjusting the space between bobbins with a screw-like mechanism.

Figure B-31 Bobbin-style descent control device.

Bolt Cutters A tool designed to make a precise, controlled cut (Figure B-32), used for cutting wire, fencing, bolts, and small bars. Also called bolt clippers, a bolt trimmer, or a rivet clipper.

Figure B-32 Bolt cutters.

Bomber See *Air Tanker*

Bonnet See *Hydrant*

Boom A floating physical barrier that serves as a continuous obstruction to the spread of a contaminant, such as an absorbent pad that collects spilled material on the water surface (Figure B-33).

Figure B-33 Floating boom.

Booster Line or Booster Hose A hose that is usually one inch in diameter and rubber-jacketed (Figure B-34), it is used on small fires in conjunction with water from an apparatus booster tank. These lines are usually stored on reels. Also called a red line.

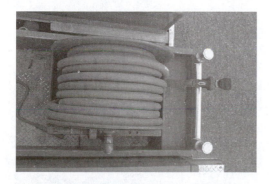

Figure B-34 Booster hose.

Booster Pump See *Pump*

Booster Reel A mounted reel on which booster hose is carried and stored (Figure B-35).

Figure B-35 Booster reel.

Booster Tank The tank on a pumper or quint that supplies booster lines and hand lines at a fire until a connection with a water source can be made. The booster tank on most pumpers holds between 500–1,000 gallons. The tank on a ladder truck with a booster tank (also known as a quint) is usually smaller, carrying only a couple hundred gallons.

Boot Protective gear worn on the feet. There are many types of boots used in the fire service today, such as bunker, crash, and wildland boots. See also *Personal Protective Equipment*

Boston Rake A tool designed to pull plaster and lath. Although it will function as a pike pole, the Boston rake is a superior tool for use in older-construction buildings that have heavy plaster and lath, wire lath, heavy baseboards and trim, and large window trims and frames (Figure B-36).

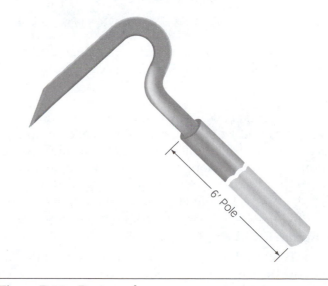

6' Pole

Figure B-36 Boston rake.

B

Bourdon Gauge   The type of gauge found on most fire apparatus that operates by pressure in a curved tube moving an indicator needle (Figure B-37). See also *Gauge*

Figure B-37   Bourdon gauges.

Bourke Eye Shield   A shield for one's upper-face that mounts on a helmet brim (Figure B-38).

Figure B-38   Bourke eye shield.

Box Fire Alarm   Allows a coded or voice message to be generated from an alarm box located in highly visible, easily accessible areas that are open to the general public. There several different types of fire alarm boxes, such as the manual pull box (Figure B-39A), the signal switch box (Figure B-39B), and the solar-powered call box (Figure B-39C).

Braid-on-Braid Rope   See *Rope*

Break Bar Rack   A versatile descent control device used by rescuers as a friction surface to control a descent by rope (Figure B-40).

Figure B-39A   Manual pull box.

Figure B-39B   Call box with signal switch.

Figure B-39C   Solar-powered call box. (Photo courtesy of Caddo Parish 9-1-1.)

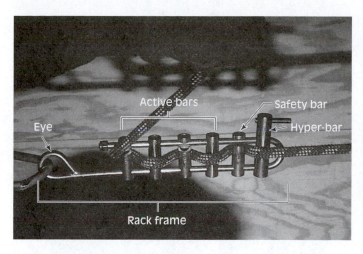

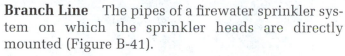

Figure B-40   Brake bar rake (open end).

Figure B-42   Brush engine or wildland engine.

Branch Line   The pipes of a firewater sprinkler system on which the sprinkler heads are directly mounted (Figure B-41).

Brass   Another term for appliances or equipment. See also *Hose Appliance*

Brush Engine   Any light vehicle with limited pumping and water capacity, designed for initial attack on small wildland fires (Figure B-42). See also *Engine Types*

Break Apart Nozzle   See *Nozzle*

Breathing Air System   See *Air Compressor*

Breathing Apparatus   Firefighters use breathing apparatus for various operations. They usually use self-contained breathing apparatus (SCBA) to enter

areas that have a hazardous or oxygen-deficient atmosphere. There are four respiratory hazardous conditions that require firefighters to wear SCBAs—oxygen deficiency, high temperature, smoke by-products, and toxic environments. All breathing apparatus used by the fire service is required to be positive pressure. There are two types of SCBAs used in the fire service today: open-circuit and closed-circuit systems. In open-circuit systems (Figure B-43A), the exhaled air is vented to the outside atmosphere. There are four components of the open-circuit breathing apparatus (Figure B-43B)—the cylinder assembly, the face piece, the regulator assembly, and the backpack harness assembly. With closed-circuit breathing apparatus (Figure B-43C),

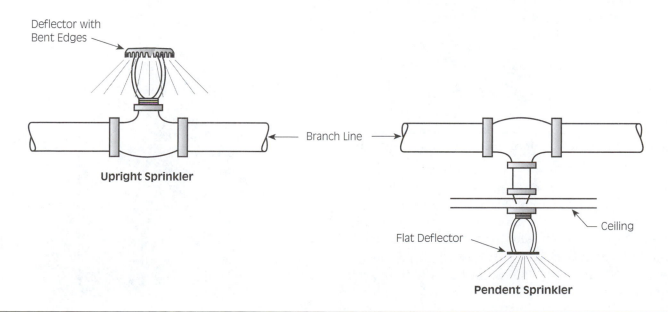

Figure B-41   Branch line.

B

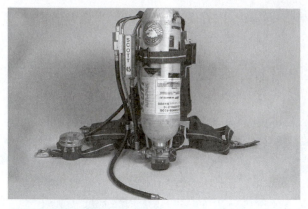

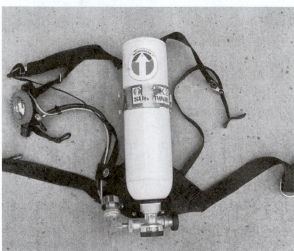

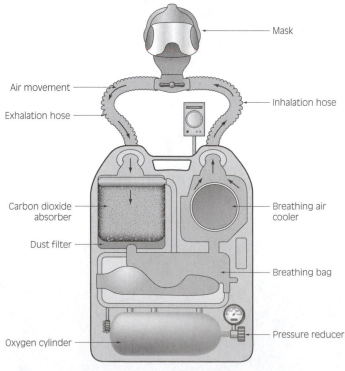

Figure B-43A Open-circuit self-contained breathing apparatus.

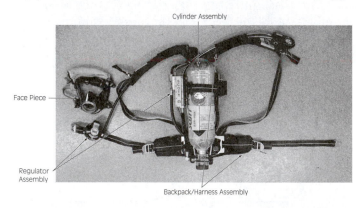

Figure B-43B Components of the open-circuit breathing apparatus.

the exhaled air stays in the system for filtering, cleaning, and circulation. The closed-circuit breathing apparatus uses a carbon dioxide absorber, an air cooler, and an oxygen tank to circulate the air.

Breathing Apparatus Air Tanks The fire service uses several different types of air tanks with the self-contained breathing apparatus (SCBA). Figure B-43D shows general descriptions of the different types of cylinders.

Figure B-43C Closed-circuit breathing apparatus.

B

Figure B-43D *From l. to r.:* Steel, aluminum, hoop-wrapped fiberglass, Kevlar composite, and fiber composite breathing apparatus cylinders.

Bresnan Distributor A nozzle with six or nine solid tips or broken-stream openings designed to rotate in a circular spray pattern, used in basements or cellars when firefighters cannot make a direct attack on the fire (Figure B-44).

Figure B-44 Bresnan distributor nozzle.

Brush Gear Another term for a Wildland Personal Protective Ensemble.

Brush Hook See *Axe*

Brush Patrol Any light vehicle with limited pumping and water capacity for off-road operations (Figure B-45). See *Engine*

Figure B-45 Brush patrol.

Bubble "B" See *Personal Protective Equipment*

Bulldozer Also called a dozer; any tracked vehicle with a front-mounted blade used to expose mineral soil (Figure B-46), usually moved by a bulldozer transport to an incident (Figure B-47). Bulldozers are rated by the National Wildfire Coordinating Group by type.

- A type-one (or heavy) bulldozer has a minimum of 200 horse power and include the D-7, D-8, or equivalent.
- A type-two (or medium) bulldozer has a minimum of 100 horse power and include the D-5, D-6, or equivalent.
- A type-three (or light) bulldozer has a minimum of 50 horse power and includes the D-4 or equivalent.

Figure B-46 Bulldozer.

Figure B-47 Bulldozer on a transport.

B

Bulldozer Tender Any ground vehicle with personnel capable of bulldozers' maintenance, minor repairs, and limited fueling.

Bumper Line A fire hose usually stored in the front bumper, pre-connected to a discharge for quick deployment in most cases (Figure B-48).

Figure B-48 Bumper line or bumper hose.

Bunkers A slang term that is used mostly to describe the components of a structure firefighting ensemble. The original term referred only to the pant/boot combination that firefighters wore at night and placed next to their "bunks" for rapid donning.

Burn Building Used to train firefighters under real conditions and effects of fire and smoke (Figure B-49). Also called a training tower.

Figure B-49 Burn building or training tower.

Burning Torch See *Back Fire Torch*

Butt The end of a ladder or hose. See also *Ladder* and *Hose*

Butterfly Valve See *Valve*

Bypass Eductor A device with two waterways and a valve that allows water to pass by or through the venturi to create a foam solution (Figure B-50).

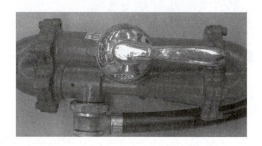

Figure B-50 Bypass eductor bypass valve.

Cab Area of an apparatus where all crews are transported, with communications, maps, computers, and sometimes GPS (Figure C-1). The National Fire Protection Association (NFPA) 1901 requires a total enclosure of all crew areas with requirements for seats and seat belts for all members.

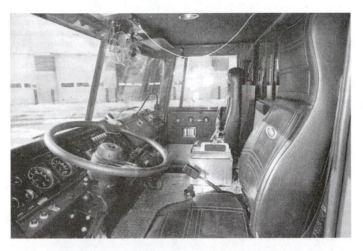

Figure C-1 Cab.

CAD Computer-aided dispatch (CAD) is a computer resource that tracks resources, the locations of active incidents, and which units have been assigned to respond to those incidents (Figure C-2).

Figure C-2 Computer-aided dispatch workstation.

Canteen Portable flask or container used by firefighters for carrying liquids such as drinking water. One of the most important survival tools a wildland firefighter can have is a canteen to prevent dehydration. Canteens come in various sizes and shapes, from bags to plastic containers (Figure C-3).

Figure C-3 Canteen.

Car Door Opening Tool Opens locked doors when inserted between the glass and weather-stripping and pushed down on the control arm or by pushing it to the rear (Figure C-4).

Figure C-4 Car door opening tool.

Carabiner An oblong metal ring with a spring-hinged side that comes in several different shapes and sizes (Figure C-5A). Firefighters and rescue workers use carabiners in rope rescue operations for such purposes as clipping a freely running rope. Figure C-5B shows a mechanical gate latch design, Figure C-5C shows the pin latch design, and Figure C-5D shows the forged steel auto-locking carabiner. Figure C-6E shows the parts of the carabiner.

Cascade System A system of filling air bottles from one tank to another. The cascade system can be fixed or mobile. Air is cascaded starting with the lowest-pressure storage cylinder in the system and slowly builds up the full cylinders on the cascade cylinders (Figures C-6A and C-6B).

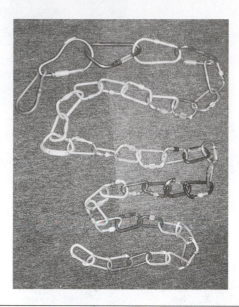

Figure C-5A Carabiners come in hundreds of different shapes and sizes.

Figure C-5B Carabiner—mechanical gate latch design.

Figure C-5C Carabiner—pin latch design.

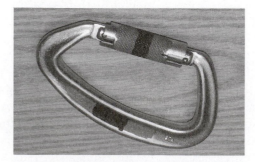

Figure C-5D Carabiner—forged steel auto-locking.

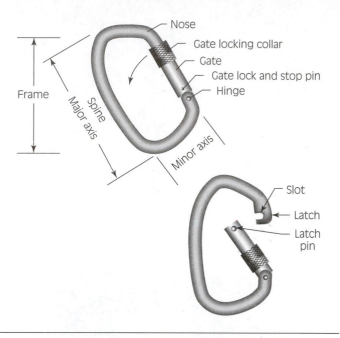

Figure C-5E Parts of a carabiner.

Figure C-6A Mobile cascade air system.

Figure C-6B Fixed cascade air system.

Catalytic Bead Sensor The most common type of combustible gas sensor uses two heated beads of metal to determine the presence of flammable gases (Figure C-7).

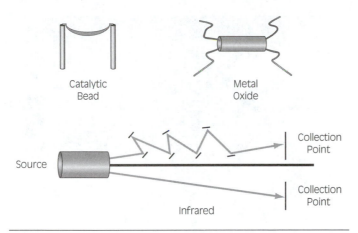

Figure C-7   Catalytic bead sensor.

Catch-All   A catch basin made from salvage covers to contain water dripping through the ceiling until a system is set up to remove the water from a building (Figure C-8).

Figure C-8   Catch-all.

Chafing Block   Placed under hose lines to protect the hose covering from rubbing against the ground or concrete (Figure C-9).

Figure C-9   Chafing block.

Chain Saw   See *Saws*

Chain Saw Chaps   Similar to the chaps used by horse riders but made of Kevlar cloth, they are designed to protect a firefighter's legs while operating a chainsaw (Figure C-10). If the saw cuts into the chaps the Kevlar abrades away, quickly jamming the saw and usually protecting the user. The chaps are generally destroyed in this process.

C

Figure C-10   Chain saw chaps.

Chalk   Firefighters use chalk to mark searched areas of a building or location during a fire or building collapse. Firefighters and rescue workers use a chalk called crayon or railroad chalk. It comes in white or blue with a 1.5-inch diameter and a 4-inch length (Figure C-11).

Figure C-11   Crayon or railroad chalk.

Charged Line   Hose filled with water that is ready for use (Figure C-12).

Figure C-12   Charged line.

Chassis The initial frame of the fire apparatus, designed to meet the department specifications set by the National Fire Protection Association (NFPA) (Figure C-13).

Figure C-13 Chassis with cab before buildup.

Chisels A tool with a cutting edge at the end of a blade used to cut, pare, or separate solid material such as metal or wood. Used by striking or by applying pressure to the end of the tool. Firefighters use chisels for heavy materials that need to be breached, to drive out rivets, to shear bolts, or to knock out heavy hinges. These chisels from the San Francisco Fire Department come in three styles—cold, backing out, and diamond point heads (Figure C-14).

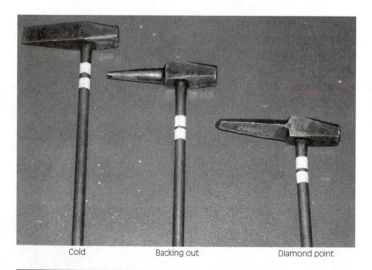

Cold   Backing out   Diamond point

Figure C-14 Chisel heads.

Chocks Wooden, plastic, or metal blocks constructed to fit the curvature of a tire, placed against the tire to prevent vehicles from rolling (Figure C-15A). Some departments attach a cable to the chock block so that it is connected to the vehicle (Figure C-15B). Chocks are also designed to grip the road to further prevent a vehicle from moving (Figure C-15C).

Figure C-15A Chock.

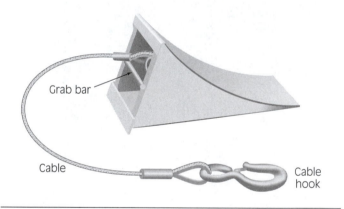

Grab bar

Cable

Cable hook

Figure C-15B Chock block with a cable that hooks under the vehicle.

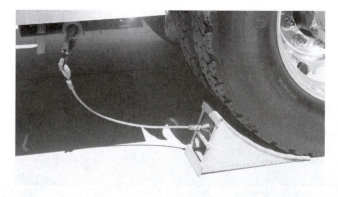

Figure C-15C Chock with cable attached to vehicle.

Chute A channel constructed of salvage covers to divert water run-off out of a building (Figure C-16).

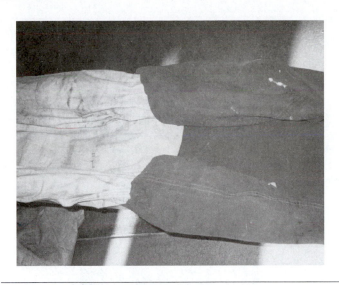

Figure C-16 Chute or water chute.

Cistern A water storage container, usually below grade level, for emergency use, including fire protection (Figure C-17).

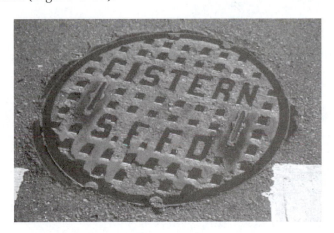

Figure C-17 Cistern.

Clinometer An optical device for measuring elevation angles above horizontal (Figure C-18A). Firefighters use clinometers to predict fire behavior and develop fire attack plans. A clinometer can tell a firefighter if the slope he is traveling is too steep, which can be indicated on a map. Some compasses have clinometers built in (Figure C-18B). There are several factors that influence the use of clinometers:

- A clear line of sight between the user and the measured point is needed.
- A well-defined object is required to obtain the maximum precision.
- The angle measurement precision and accuracy is limited to slightly better than one degree of arc.

Simple clinometers can be made with a protractor, string, and a weight (Figure C-18C).

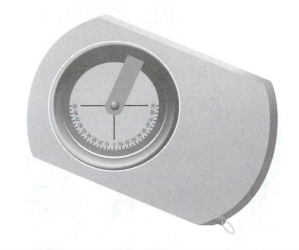

Figure C-19A Clinometer.

Figure C-19B Clinometer on a compass.

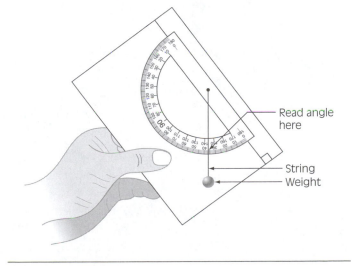

Read angle here

String
Weight

Figure C-19C Simple clinometer.

Clipboard Firefighters use clipboards for paperwork in the field. The clipboard used by this firefighter is a sturdy box that can also carry the paperwork out of the elements (Figure C-19).

Figure C-19 Firefighter using a clipboard that is built to withstand harsh conditions.

Closed-In Rescue Rack A rope rescue lowering system with a closed-in rigging that applies friction to a rope and is controlled by changing the angle of approach the rope has to the rack (Figures C-20A, C-20B, and C-20C).

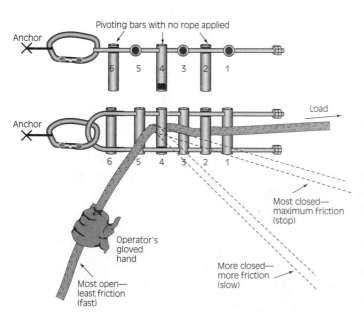

Figure C-20A Closed-in rescue rack: the system.

Figure C-20B All bars activated.

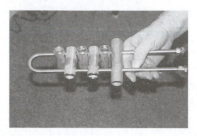

Figure C-20C The more bars are added, the more friction available.

Collapsible Firefighting Rake Used by firefighters for line construction. Made of stainless-steel tines and designed to collapse (Figure C-21).

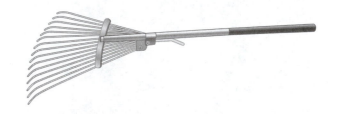

Figure C-21 Collapsible fire rake.

Collapsible Tank A water tank made of rubber or synthetic material that collapses for easy storage (Figure C-22).

Figure C-22 Collapsible tank.

Colorimetric Tube Glass tubes containing a chemically treated substance that reacts with specific airborne chemicals to produce a distinctive color. The tubes are calibrated to indicate approximate chemical concentrations in the air (Figure C-23).

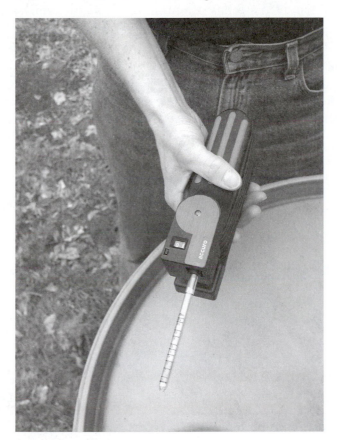

Figure C-23   Colorimetric tube.

Command Center A department communication office where decisions are made and from which resources are dispatched (Figure C-24), also known as a dispatch center or communications center.

Figure C-24   Command center.

Command Vehicle A vehicle used by operations chief officers (Figure C-25).

Figure C-25   Command vehicle.

Combination Headset A device that protects the user's hearing and also enables communication between crew members and other responders (Figure C-26).

Figure C-26   Combination headset and hearing protection.

Combination Tool The jack of all trades of fire tools, the combination or combi tool is basically an enlarged version of the military entrenching tool. It has a long handle (approximately 5 feet) and a small shovel head, which can be folded out straight with the handle to be used as a shovel or it can be folded at 90 degrees to the handle for use as a hoe or scraping tool. The back side has a pick-like point, which can also be folded inline with the handle or at a 90-degree angle. It is often referred to as the crew boss or officer's tool (Figure C-27).

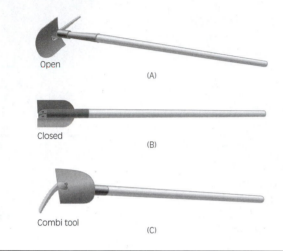

Open (A)

Closed (B)

Combi tool (C)

Figure C-27 Combination tool or combi tool.

Compass An instrument that indicates magnetic north consisting of a strip of magnetized steel balanced on a pinpoint and free to swing in any direction. One end of the needle, usually marked by an arrow or luminous red dot, is the north-pointing end. When the compass is laid or held horizontally, the red end of the needle will always point north to magnetic north. The compass consists of three basic parts (Figure C-28):

1. The magnetic needle that points to magnetic north

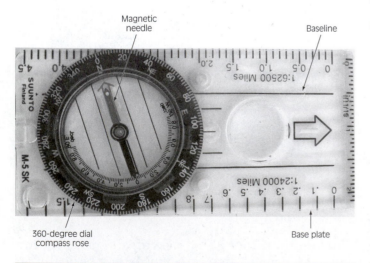

Magnetic needle

Baseline

360-degree dial compass rose

Base plate

Figure C-28 Parts of a compass.

2. A revolving 360-degree dial that is a graduated circle which shows north, east, south, and west, sometimes divided into four or twelve additional points called the *compass rose*. The dial's upper rim has degree indentations, and the lower transparent plate has orienting north/south lines and an arrow.

3. A transparent "base plate" that has direction and ruled edges.

Compass Rose A circle, graduated in degrees, printed on some charts or marked on the ground at an airport or heliport, used as a reference to either true or magnetic direction. The compass rose is the degree dial on the compass (Figure C-29).

Figure C-29 Compass rose.

Compound Gauge A pressure gauge on fire department engines that records the pressure above and below atmospheric pressure (Figure C-30). The pressure below atmospheric pressure is usually measured in inches of mercury and is called *vacuum*. The pressure above atmospheric pressure is measured in pounds per square inch or kilopascals.

Figure C-30 Compound gauge.

Compressed Air Foam System (CAFS)

Compressed air is injected into the foam solution prior to entering any hose lines. The fluffy foam created needs no further aspiration of air by the nozzle (Figure C-31).

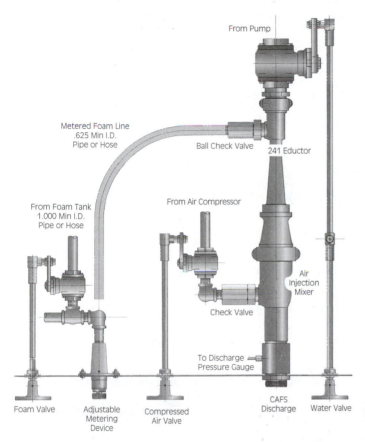

Figure C-31   Compressed air foam system (CAFS).

Compressor Purifier

A compressor system for filling breathing apparatus bottles (Figure C-32).

Figure C-32   Compressor purifier.

Communication Unit

Usually a van, mobile trailer, or bus used to provide an incident with communication capabilities (Figure C-33).

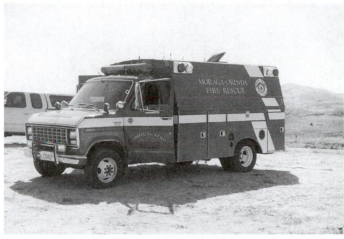

Figure C-33   Communication unit or van.

Combustible Gas Detector

An instrument designed to detect the presence or concentration of combustible gases or vapors in the atmosphere (Figure C-34).

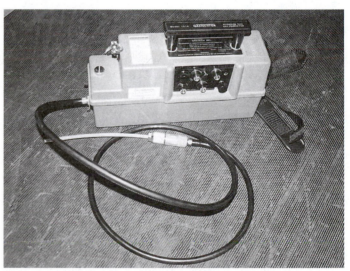

Figure C-34   Combustible gas detector.

Control Panel

The location where firefighters go to check a building's heat and smoke detection system (Figure C-35).

Coordinate Ruler

A map tool with a set of scales printed on clear plastic. These scales are used like a ruler to measure different coordinate systems. The scales can be used to measure distance in feet and to determine slope (Figure C-36).

Figure C-35 Firefighter checking fire alarm system control panel.

Council Rake Common in the southern and eastern United States, this tool resembles a garden rake that has had its tines replaced with metal shark teeth. It is a scraping tool used to cut through thick layers of leaf or needle litter while making fire line. The sharp teeth are also useful for cutting vines. Also called a council tool, rich tool, or fire rake (Figure C-37).

Figure C-37 Council rake.

Coupling See *Hose Appliances*

Crash Fire Rescue (CFR) Older term for airport aircraft rescue and firefighting (ARFF) resources. See *Airport Crash Truck*

Crawler Tractor A tracked vehicle often equipped with a front-mounted blade and rear-attached fire plow used to suppress wildfires (Figure C-38).

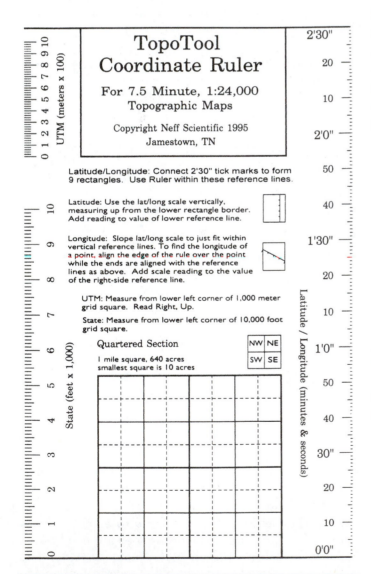

TopoTool Coordinate Ruler

For 7.5 Minute, 1:24,000 Topographic Maps

Copyright Neff Scientific 1995
Jamestown, TN

Latitude/Longitude: Connect 2'30" tick marks to form 9 rectangles. Use Ruler within these reference lines.

Latitude: Use the lat/long scale vertically, measuring up from the lower rectangle border. Add reading to value of lower reference line.

Longitude: Slope lat/long scale to just fit within vertical reference lines. To find the longitude of a point, align the edge of the rule over the point while the ends are aligned with the reference lines as above. Add scale reading to the value of the right-side reference line.

UTM: Measure from lower left corner of 1,000 meter grid square. Read Right, Up.

State: Measure from lower left corner of 10,000 foot grid square.

Quartered Section

1 mile square, 640 acres
smallest square is 10 acres

| | NW | NE |
| | SW | SE |

Figure C-36 Coordinate ruler.

Figure C-38 Crawler tractor.

Crew An organized group of firefighters under the leadership of a supervisor (Figure C-39).

- **Helitack Crew** A crew that performs helicopter management and attack activities.

- **Hand Crew** A number of individuals who have been organized and trained and are supervised principally for operational assignments on an incident, also called fire crews, (Figure C-40A). Fire lines cut by hand crews average 3 feet in width in grass to more than 15 feet in heavy fuels. Figure C-40B shows a hand crew cutting line at a wildland fire. The crew standards for national

Figure C-39 Crew.

Figure C-40A Hand crew.

Figure C-40B Hand crew cutting fireline.

mobilization according to the National Wildfire Coordinating Group are as follows:

Type-One Crew

18–20 members
80% of crew with one or more seasons of experience
80 hours of training
Arduous fitness standard
Own transportation
Fully equipped with tools and equipment
Self-sufficient
Dispatch available within one hour
Maximum weight of 5,100 lbs.
Full-time organized crew
3 agency qualified sawyers
5 portable programmable radios
Permanent supervision

Type-Two Crew

18–20 members
60% of crew with one or more seasons of experience
Basic firefighter training
Arduous fitness standard
Transportation needed
Not fully equipped
Not self-sufficient
Dispatch available is variable
Maximum weight of 5,100 lbs.
Not a full-time organized crew
3 agency qualified sawyers
4 portable programmable radios

Note: A Type-Two IA (initial attack) can be broken into squads and a Type-Two non-IA cannot. A Type-Two IA requires 60% of its crew to have one season's experience and a non-IA Type-Two crew requires 40% of the crew to have one season or more of experience. A Type-Two none IA does not require qualified sawyers.

Type-Three Crew

18–20 members
40% of crew with one or more seasons of experience
Basic firefighter training
Arduous fitness standard
Transportation needed
Not fully equipped
Not self-sufficient
Dispatch available is variable
Maximum weight of 5,100 lbs.
Not a full-time organized crew
4 portable radios

C

- **Hotshot Crew** Intensively trained fire crew used primarily in hand line construction (Type one). Interagency hotshot crews (IHC) are a Type-one crew that exceeds the Type-one standards as required by the National IHC Guide (2001) in the following categories:
  - Have a permanent supervisor with seven years of full-time appointments (superintendent, assistant superintendent, 3 squad bosses)
  - Work and train as a unit 40 hours per week
  - Be available as a national resource

Crew Carrying Vehicle Used to transport hand crews; also called a crew bus, crew van, or crew carrier (Figure C-41).

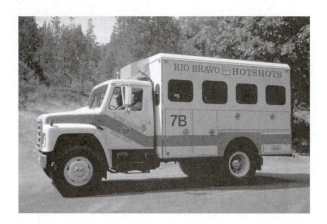

Figure C-41 Crew bus or crew carrying vehicle.

Cribbing Varying lengths of hardwood, usually 4 × 4 inches or larger, used to stabilize objects, particularly vehicles during extrication incidents (Figure C-42). See also *Blocks*

Figure C-42 Cribbing being use to stabilize a vehicle.

Crotch Pole A device attached to a tormentor or 50-foot pole ladder for downhill operations (Figure C-43).

Figure C-43 Crotch pole—an extension for a 50-foot pole ladder.

Cutter Hydraulic-operated rescue tool that can cut through metal with a force up to 70,000 lbs. (Figure C-44A). Primarily used in automobile extrications to cut vehicle posts, but can be used in a variety of applications. Figure C-44B shows a steering column being severed by hydraulic cutters. There are other types of cutters used in the fire service today.

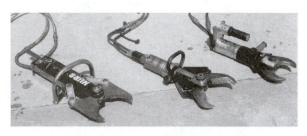

Figure C-44A Power hydraulic cutters. (Photo courtesy of Rick Michalo.)

Figure C-44B Cutters being used to seven a steering column. (Photo courtesy of Rick Michalo.)

- **Cable Cutter** Similar to the bolt cutter but with a more rounded cutting area designed to cut cables or wire rope (Figure C-45).

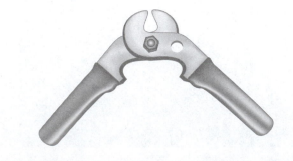

Figure C-45 Cable cutters.

- **Ring Cutter**   A device for removing rings by cutting (Figure C-46).

Figure C-46   Ring cutter.

- **Wire Steck Cutter**   An older style of electrical wire cutter with rubber handles to prevent shock (Figure C-47).

Figure C-47   Steck wire cutters.

Cutting Torch   A device that uses a mixture of acetylene and oxygen to generate a high-temperature flame for cutting metal by melting it (Figure C-48); used for rescue and forcible entry.

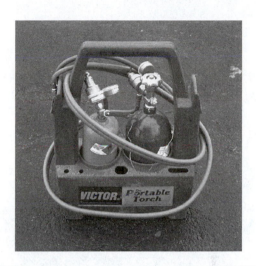

Figure C-48   Cutting torch.

Cylinders   See *Air Breathing Apparatus Bottles*

D-Handle   A D-shaped handle that is used on fire-fighting tools such as poles, hooks, and some forcible entry tools (Figure D-1).

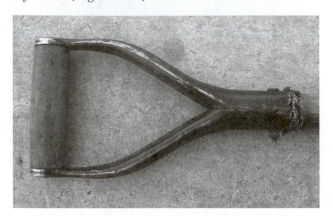

Figure D-1   D-handle.

D-Ring   A D-shaped ring that is used to attach a rope or equipment to other devices (Figure D-2A).

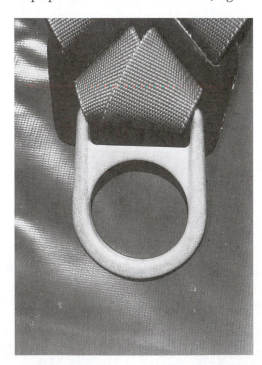

Figure D-2A   D-ring.

Shown here is a D-ring attached to a Class-Three harness (Figure D-2B).

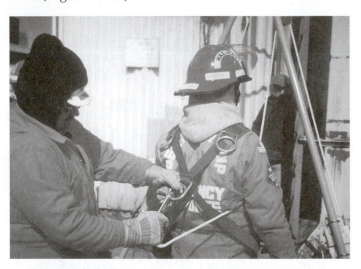

Figure D-2B   D-ring attached to a Class-Three harness.

Day Pack   A backpack used by firefighters to carry personal belongings, fire tools, and equipment (Figure D-3).

Figure D-3   Day pack.

Debris Carrier   Heavy canvas, usually 4 feet square, with hand straps at each corner. Firefighters use this to remove debris during overhaul operations (Figure D-4). Also called a carry-all; can be used to carry large patients.

Figure D-4 Debris carrier.

Figure D-5B Simple decontamination shower.

Deck Gun A portable, heavy-stream application device used for firefighting. It is set up on the ground at a fire scene and supplied by large-diameter hoses for the application of large quantities of water onto a fire. Also known as a deck pipe.

Decontamination Showers A portable shower used for removal of contaminants from people and equipment. There are several types of decontamination showers—from those constructed of synthetic materials (Figure D-5A) to those that are just a single fire hose and a salvage cover (D-5B).

Delayed Aerial Ignition Device (DAID) See *Aerial Ignition Device*

Deluge Set A large-capacity firewater nozzle classification. It generally consists of a short length of

large-diameter hose having a large nozzle with a tripod support at one end and a three- or four-way Siamese or inlet connection at the supply end.

Deluge System A sprinkler system using open sprinklers installed in a piping system connected to a water supply through a valve that is opened by the operation of a redetection system installed in the same area as the sprinklers. In a deluge system, all sprinklers are activated.

Denver Tool A combination maul, axe prying, and pulling tool (Figure D-6). Also called the TNT Tool.

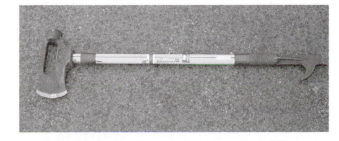

Figure D-6 Denver tool.

Descent Control Device (DCD) A friction-producing device that rubs a rope and slows its movement. There are several types of descent control devices (Figure D-7A) such as the carabiner (Figure D-7B), the single tube DCD (Figure D-7C), the O-ring (Figure D-7D), the eight plate (Figure D-7E), and the brake bar rack (Figure D-7F).

Figure D-5A Manufactured decontamination shower.

D

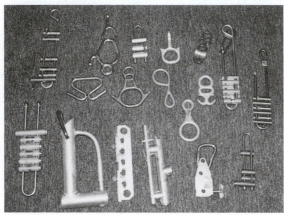

Figure D-7A   Various descent control devices.

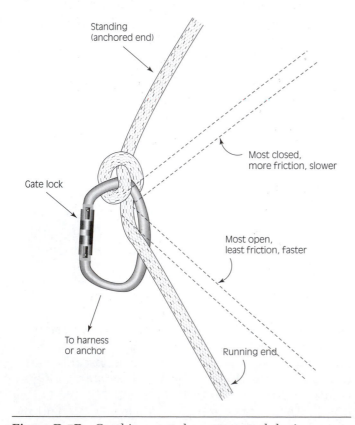

Figure D-7B   Carabiner as a descent control device.

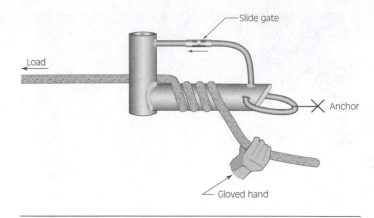

Figure D-7C   Single tube DCD.

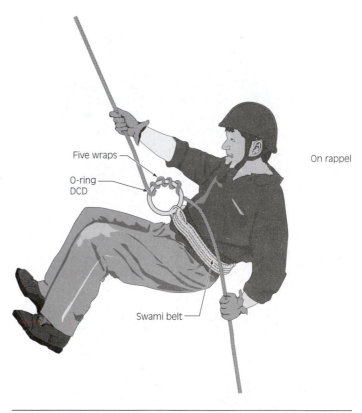

Figure D-7D   An O-ring being used as a descent control device.

Figure D-7E   Eight plate being used as a DCD.

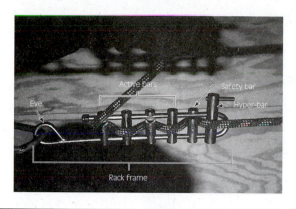

Figure D-7F   Brake bar rack (open end).

Detectors   A detection system is designed to notify people that a potentially life-endangering event is happening (Figure D-8A). The following are different types of detectors:

- *Rate-of-Rise Detector* Measures temperature increase above a predetermined rate (Figure D-8B).

- *Fixed-Temperature Heat Detector* Usually electrically operated with a bimetallic strip that expands at different rates (Figure D-8C).

- *Ionization-Type Smoke Detector* The most common type of smoke detector, using a radioactive element that emits ions into a chamber. The positive and negative ions are measured on an elec-

trically charged electrode. When smoke enters the chamber, the flow of ions to the electrode changes and an alarm is activated (Figure D-8D).

- *Photo-Electric Detectors* Use two different methods to detect the smoke and can be spot or line detectors. The first is the light-obscuring detector, which has a light beam at a light sensor (Figure D-8E). When smoke particles obscure or block the light beam, the light sensor notes the loss of light and activates the alarm. The light-scattering detector operates by a light beam aimed at the end of the chamber with a light sensor in an angled-off chamber. Smoke particles scatter the light, which strikes the light sensor, activating the alarm (Figure D-8F).

- *Combustible Gas Indicator (CGI) Detector* Measures the presence of a combustible gas or vapor in the air.

- *Photo-Ionization Detector (PID)* Used to determine the presence of gases/vapors in low concentrations in the air.

- *Gas Chromatograph/Mass Spectrometer Detector* Used for identifying and analyzing organics.

- *Corrosivity (pH) Detector* A meter or paper that indicates the relative acidity or alkalinity of a substance, generally using an international scale of 0 (acid) through 14 (alkali-caustic).

Figure D-8A   Smoke and carbon monoxide detectors are available separately and as one unit.

D

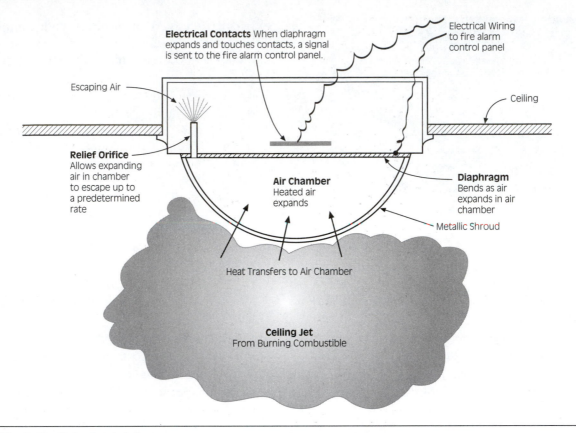

Figure D-8B   Rate-of-rise heat detector.

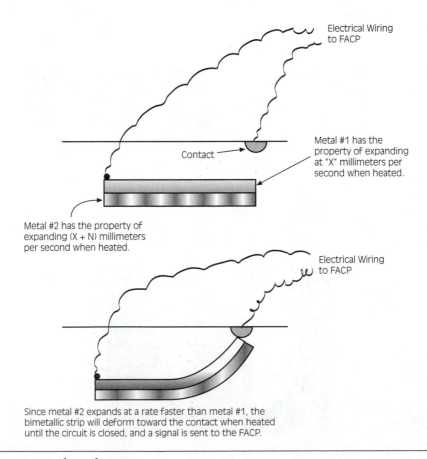

Figure D-8C   Fixed-temperature heat detector.

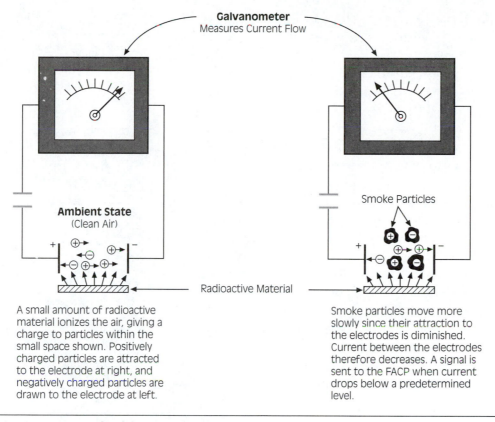

A small amount of radioactive material ionizes the air, giving a charge to particles within the small space shown. Positively charged particles are attracted to the electrode at right, and negatively charged particles are drawn to the electrode at left.

Smoke particles move more slowly since their attraction to the electrodes is diminished. Current between the electrodes therefore decreases. A signal is sent to the FACP when current drops below a predetermined level.

Figure D-8D Ionization-type smoke detector.

Light-Scattering Type

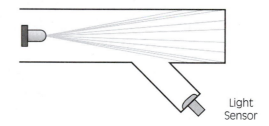

Typical Photoelectric Detector Chamber with Clean Air

Light Obscuration Type

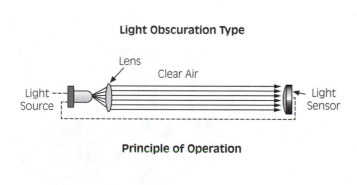

Principle of Operation

Light Obscuration Type

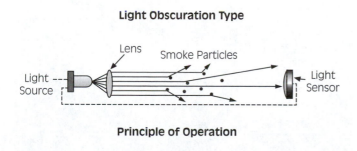

Principle of Operation

Light-Scattering Type

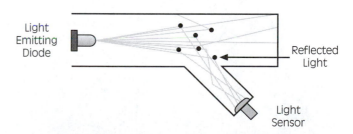

Typical Photoelectric Detector Chamber with Smoke

Figure D-8E Light obscuration photo-electric smoke detector.

Figure D-8F Light-scattering photo-electric smoke detector.

D

- *Flame Ionization Detector (FID)* Used to determine the presence of hydrocarbons in the air.
- *Ultraviolet Flame Detector* Senses the radiation of high-intensity flaming fires by detecting the light waves emitted in the UV spectrum.
- *Radiation Beta Survey Detector* Used to detect beta radiation.
- *Radiation Dosimeter Detector* Measures the amount of radiation to which a person has been exposed.
- *Radiation Gamma Survey Detector* Used for the detection of ionizing radiation, principally gamma radiation, by means of a gas-filled tube.
- *Temperature Detector* An instrument, either mechanical or electronic, used to determine the temperature of ambient air, liquids, or surfaces.

Detroit Door Opener   A two-piece pry bar used for forcing doors (Figure D-9A). Figure D-9B shows a firefighter using the tool to open a door.

Digital Recording Device   Records all telephone and radio traffic at communications centers (Figure D-10).

Figure D-9A   Detroit door opener.

Figure D-10   Digital recording device.

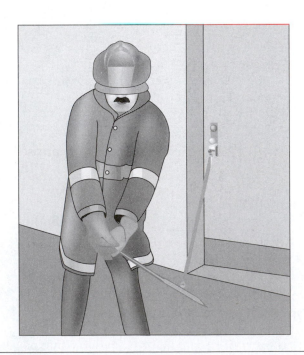

Figure D-9B   Firefighter using a Detroit door opener.

Dike   An embankment or ridge made to control the movement of liquids, sludge, solids, or other materials. Dikes are made of synthetic materials that absorb hazardous liquids or stop the flow of liquids (Figure D-11).

Figure D-11   Dike.

Direct Injection Foam The direct injection method uses a separate positive-pressure pump to inject the foam concentrate into the water pump's discharge. The correct rate of concentrate is controlled by a microprocessor, which receives an electrical signal from a flow meter attached to the discharge and adjusts the injection rate as needed (Figure D-12).

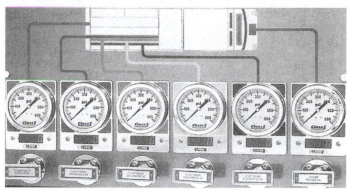

Figure D-14 Discharge gauge.

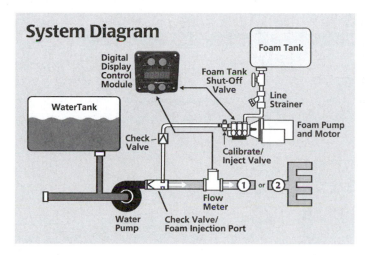

Figure D-12 Direct inject foam systems.

Direct Reading Instrument A detection and monitoring instrument that gives a reading based on a graduated scale (Figure D-13).

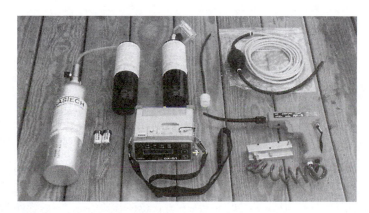

Figure D-13 Direct reading instrument.

Discharge Gauge A gauge at the pump panel that shows the pressure being discharged out of the pump by pounds per square inch (Figure D-14).

Dispatch Center A location that receives reports of discovery and status of fires and other emergencies, confirms their locations, takes action promptly to provide people and equipment likely to be needed for control in first attack, and sends them to the proper place (Figure D-15).

Figure D-15 Dispatch center.

Dispatch Center Radio System There are several systems used today to communicate with the resources in the field.

- **Dispatch Center Repeater System** A duplex design with multiple transmitters extending the operating range of the simplex system by using repeaters (Figure D-15A).
- **Dispatch Multi-Site Trunked Radio System** Extends beyond the normal range of standard repeater systems by linking multiple repeaters together (Figure D-15B).
- **Dispatch Simplex Radio System** Inexpensive but with limited range of operation (Figure D-15C).

D

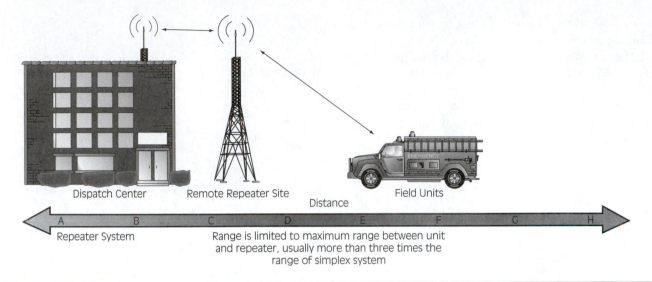

Figure D-15A Dispatch center repeater system.

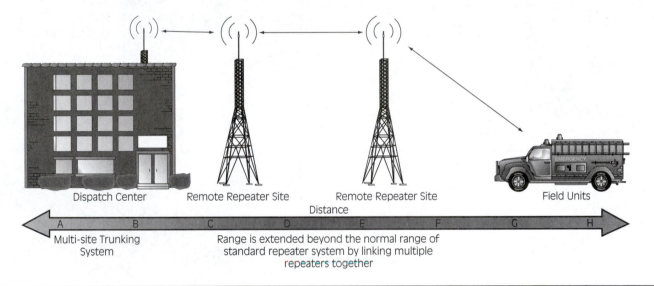

Figure D-15B Dispatch center multi-site trunking system.

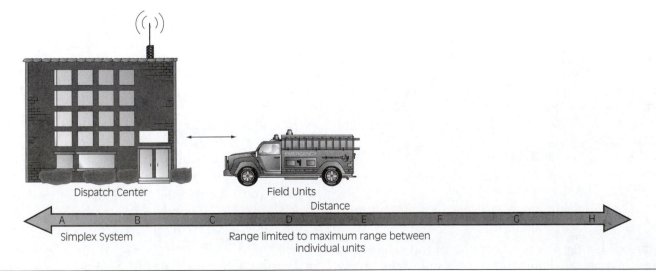

Figure D-15C Dispatch center simplex system.

D

Dispatch Sheet A card that the dispatcher uses to collect information such as person reporting, incident type, location of incident, and call back number (Figure D-16).

Door Marker A highly visible flat piece of synthetic or rubber material that is hung on a door to indicate that a search has been completed (Figure D-17).

Town of Colonie **FIRE/EMS** Incident		DATE _____			RUN #_____	
How Received : [ ] TELEPHONE [ ] RADIO [ ] IN PERSON [ ] 2ND/3RD PARTY						
Person Reporting:		Incident/Priority Type:				
Address:		Dept's				
Location of Incident:		Units				
Cross Street/Landmark:						
Callback #:		Rec'd				
Chief Complaint:		Disp'd				
Age:	Sex: Male / Female	Ack'd				
Conscious: Y / N / Unknown		Enrt				
Breathing: Y / N / Unknown		Arrv'd				
Rec'd By:	Disp By:	I/S				
EMS Zone:	Fire Zone:	Amb:	Hosp:	Enrt:	Arrv:	
Fire Signal/Time: /		Amb:	Hosp:	Enrt:	Arrv:	
Priority Change/Time: ____ / ____	Remarks: _____					

Mutual Aid Dept/Unit	Disp'd	Ack'd	Enrt	Scene	I/S

Amb:	Hosp:	Enrt:	Arrv:	Remarks:
Amb:	Hosp:	Enrt:	Arrv:	

Sig	Cover Co.	Comm. Post	Fire Inv	Op. Off	Comm. Dir.	Mob. Air	Batt. Co	Car 1
30								
40								

*Signal 20 gas leaks or hazmat incident: page fire investigation (399)

Figure D-16 Dispatch sheet.

Figure D-17 Door marker.

Door Wedge A small wedge-shaped piece of wood or synthetic material that is placed under a door to keep it open (Figure D-18).

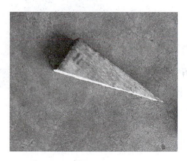

Figure D-18 Door wedge.

Distributor Nozzle A firefighting spray nozzle normally used to combat basement-level fires. Also called a Cellar Pipe. See also *Nozzle*

Distributor Pipe A device that allows a nozzle or other device to be directed into a hole to reach basements, attic, and floors that cannot be accessed by personnel. The distributor pipe has self-supporting brackets that help hold it in place when in use. Also called an extension pipe. See also *Nozzle*

Dogs See *Ladder*

DOT Emergency Response Guide A book that is published almost every three years that provides information regarding the potential hazards of materials. It is one of the only books that provides specific evacuation recommendations. The Department of Transportation makes a copy available for every emergency response apparatus in the country (Figure D-19). The book consists of these major sections:

- Placard information
- Listing by DOT identification number
- Alphabetical listing by shipping name
- Response guides

Figure D-19 DOT emergency response guidebook.

- Table of initial isolation and protective action distance
- List of dangerous water-reactive materials

Dozer See *Bulldozer*

Drafting Pit An underground reservoir of water usually located at a training center from which to draft for engine testing (Figure D-20).

Figure D-20 Drafting pit.

Dragon Slayer A tool used to rip and cut gypsum board as well as plaster and lath (Figure D-21).

Figure D-21 Dragon slayer.

Drain Protector A pad that is placed over manholes and large drains to keep hazardous materials or unwanted run-off out of the drains (Figure D-22).

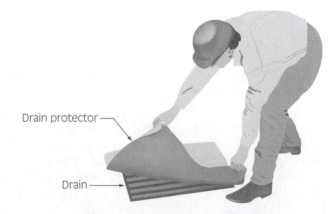

Drain protector

Drain

Figure D-22 Drain protector.

Drill Tower A training structure normally more than three stories used by training personnel to develop realistic fire service situations (Figure D-23).

Figure D-23 Drill tower.

Drip Torch Hand-held device for igniting fires (Figure D-24A) by dripping flaming liquid fuel on the materials to be burned. Figure D-24B shows the drip torch in the ready position and Figure D-24C shows

Figure D-24B Drip torch in the ready-to-use position.

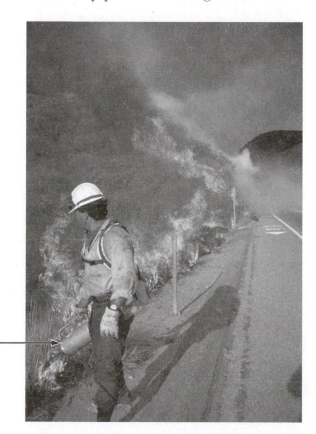

Figure D-24A Firefighter using a drip torch.

Figure D-24C Drip torch in storage position.

the drip torch in the storage position with a bracket for travel. Depending on use, fuel is generally a mixture of two-thirds diesel and one-third gasoline.

Drop-In Unit A removable firefighting pump and tank that can be loaded onto the back of a truck. Sometimes called slip-on units (Figure D-25).

Figure D-25 Drop-in unit, sometimes called a slip-on unit.

Dry Bulb Thermometer In a psychrometer, the (dry bulb thermometer) is not covered with muslin; used to determine air temperature (Figure D-26).

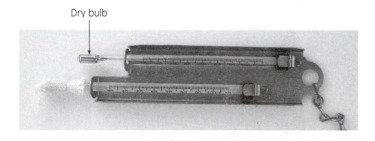

Dry bulb

Figure D-26 A dry bulb thermometer on a psychrometer.

Dry Sprinkler A sprinkler system that uses air under pressure (in place of water) until water is needed in areas that may freeze.

Dry Suit Made of waterproof materials that keep a rescue diver completely dry (Figure D-27).

Figure D-27 Dry suit.

Ducting A neoprene, wire-reinforced, flexible, fireproofed cotton duct used to carry hazardous or smoke-filled atmosphere through uncontaminated rooms to outside areas without contaminating clean rooms. Also used to carry clean air in such as in a confined space rescue operation (Figure D-28).

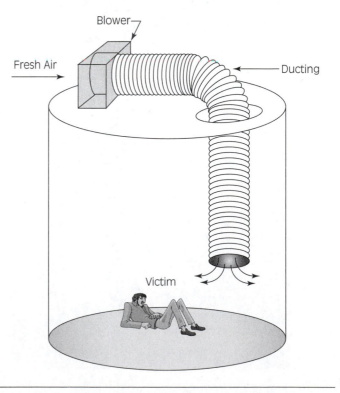

Blower

Fresh Air

Ducting

Victim

Figure D-28 Ducting attached to blower to bring fresh air in.

Dump Site The area where tenders are unloading or have dumped their load (Figure D-29).

Figure D-29 Dump site.

Dust Mask Light-weight mask with a nose clip made of paper products. Not for fighting fires but to keep dust or organic-level nuisance out of the airway (Figure D-30). Also called a particulate respirator.

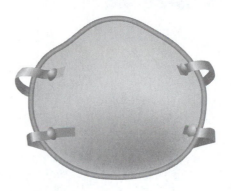

Figure D-30 Dust mask.

D

Eductor   A device that by the rapid jetting of a liquid or gas causes additional liquid or gas to be moved or picked up, such as a foam eductor (Figure E-1).

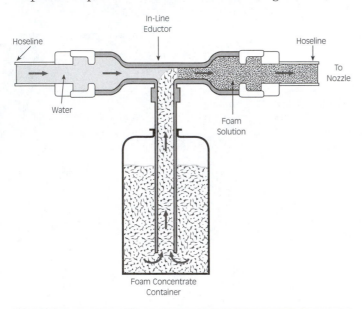

Figure E-1   Eductor.

Eight Plate   A device designed like a figure eight that is used in rope rescue work as a descent control device; also called the figure eight. There is the eight plate with ears (Figure E-2A) and the straight eight (Figure E-2B).

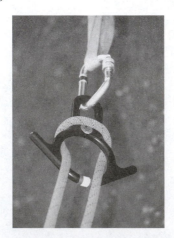

Figure E-2A   Eight plate with ears.

Figure E-2B   Straight eight.

Ejector   Also known as a hydraulic ejector or a suction booster; a jet siphon device usually supplied from a tank on a fire apparatus that is used to bring water to an engin68e from greater distances and to higher elevations than is possible with suction, depending upon atmospheric pressure (Figure E-3A).

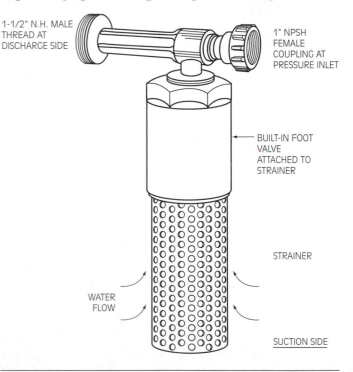

Figure E-3A   Ejector.

• Water is pumped in the conventional manner from an engine with booster line to the water inlet of the ejector, and it passes through a nozzle restriction as a high-velocity stream. This jet

E

stream then picks up additional water through the suction part and delivers the combined flow through the diffuse chamber and out the discharge port under low pressure through a 1.5-inch line.

- An ejector is very useful if the water source for drafting is 18 feet or more vertically below the pump or if the water source is farther away than the length of the draft hose. With the use of an ejector, water can be drafted and raised vertically from 40 to 250 feet, depending on the type of pump and ejector used. The ejector can be used up to several hundred feet from the pump where the pump cannot be spotted within the length of the drafting hose (Figure E-3B).

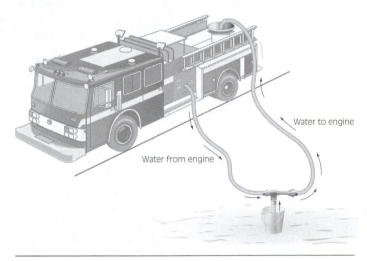

Figure E-3B   Hydraulic water ejector in use.

EK Hook *or* **Eckert Hook**   See *Hook*

Elevated Boom   The elevating mechanism of two or more booms on articulating aerial tower apparatus (Figure E-4).

Figure E-4   Elevated boom.

Elevated Platform   A hydraulically raised platform mounted on fire apparatus, designed for rescue and firefighting (Figure E-5).

Figure E-5   Elevated platform.

Elevated Water Tower   An elevated master stream mounted on an aerial apparatus that can be put into service quickly (Figure E-6).

Figure E-6   Elevated water tower.

Elevated Nozzle   See *Nozzle*

Elevator Key Box   A box of keys used to open elevator hoist way doors (Figure E-7).

Emergency Call Box   See *Box*

Emergency Egress Improvised Harness   Made from 18 to 25 feet of 2-inch tubular webbing. It is

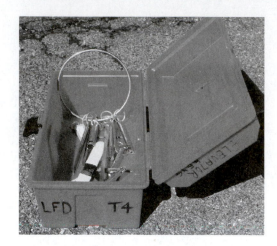

Figure E-7 Elevator key box.

specially designed to be put on in a hurry, in less than one minute with practice, and can be carried, rolled, or daisy-chained in a coat pocket for quick access (Figure E-8A). The modified Swiss Seat

improvised harness is a safer webbing harness than the emergency egress but takes longer to tie, about 3 to 5 minutes (Figure E-8B).

Emergency Response Guidebook See *DOT Emergency Response Guidebook.*

Encapsulated Suit See *Personal Protective Equipment.*

Encoder A device that converts and "enters" code into paging codes, which in turn activate a variety of paging devices (Figure E-9).

Engine A fire department pumper, also referred to as a pumper because of its ability to pump water. In most cases, an engine carries small ground ladders, supply line to connect it with a hydrant, hand lines with which to fight the fire, and a tank holding between 500 and 1,000 gallons of water (Figure E-10A).

There are several different descriptions of engines and pumpers. The engine company is organized to provide firefighters who deliver water at the fire scene, deploy hose lines, and attack and extinguish

1 This harness is tied by finding the middle of the length of webbing, which should be marked at mid-length for quick identification.

2 A bright is formed in the middle and brought up between the legs.

3 The two long tails separate behind the buttocks, surround the thighs, and are passed through the center bight.

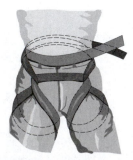

4 Each tail is then passed toward the back of the hips, crossed, and brought to one side for joining with a square bend.

5 Half hitches safe the square bend and hold it in position.

6 The webbing should always be flat and tight against the body, and the harness should be applied to the pelvic girdle with tension on the hip points and not the abdomen.

Figure E-8A Emergency egress improvised harness made from webbing.

E

1 Find a spot near the middle of the 25-feet webbing. (One side will need to be about 2 feet longer.)

2 Measure and then tie leg loops using an overhand knot.

3 Space between the leg loops should be about 3 or 4 inches only.

4 Wrap the running ends around the waist. At the first pass, make a square bend.

Square bend

5 Finish by wrapping excess webbing (if any) multiple times around hips (swami style).

6 Conclude with a square bend and overhand knot safeties on both sides of the bend.

Figure E-8B   Modified Swiss seat.

Figure E-9   Encoder.

fires. The National Fire Protection Association (NFPA) states that a pumper should have a permanently mounted fire pump with a capacity of at least 750 gallons per minute (GPM). NFPA also states that a pumper should carry no less than 300 gallons of water (Figure E-10B).

The National Wildfire Coordinating Group (NWCG) describes fire equipment as types. NWCG has seven basic types of engines—a Type-One engine

Figure E-10A   The engine company provides the firefighters with the tools, equipment, and hose necessary for fire suppression operations.

Figure E-10B An engine that meets National Fire Protection Association (NFPA) standards.

with the largest capacity and a Type-Seven engine with the least capacity. Each type is broken down into pump rating, tank capacity, hose amount and size, ladders, master stream in gallons per minute (GPM), and the minimum amount of personnel on the engine.

- Type-One Engine (Figure E-10C).
    Pump rating: 1,000+ GPM at 150 PSI
    Tank capacity: 400+ gallons
    2.5-inch hose: 1,220 feet

1.5-inch hose: 400 feet
1-inch hose: none
Ladders: 48 feet
Master stream: 500 GPM
Personnel: 4 minimum
- Type-Two Engine (Figure E-10D)
    Pump rating: 250+ GPM at 150 PSI
    Tank capacity: 400+ gallons
    2.5-inch hose: 1,000+ feet
    1.5-inch hose: 500 feet
    Ladders: 48 feet
    Personnel: 3 minimum
- Type-Three Engine (Figure E-10E)
    Pump rating: 150+ GPM at 250 PSI
    Tank capacity: 500+ gallons
    1.5-inch hose: 500 feet
    1-inch hose: 500 feet
    Personnel: 3 minimum
- Type-Four Engine (Figure E-10F)
    Pump rating: 50+ GPM at 100 PSI
    Tank capacity: 750+ gallons
    1.5-inch hose: 300 feet
    1-inch hose: 300 feet
    Personnel: 2 minimum
- Type-Five Engine (Figure E-10G)
    Pump rating: 50+ GPM at 100 PSI
    Tank capacity: 400 to 750 gallons
    1.5-inch hose: 300 feet
    1-inch hose: 300 feet
    Personnel: 2 minimum

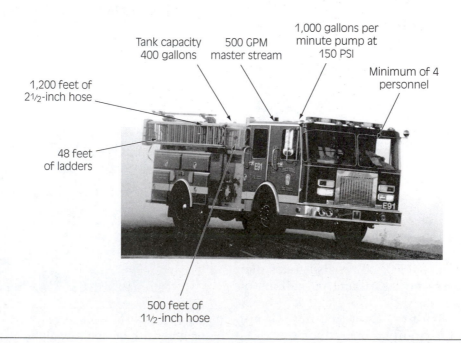

Figure E-10C Type-One engine.

E

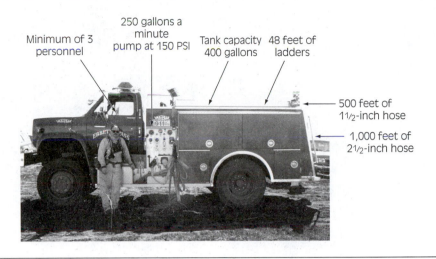

Figure E-10D   Type-Two engine.

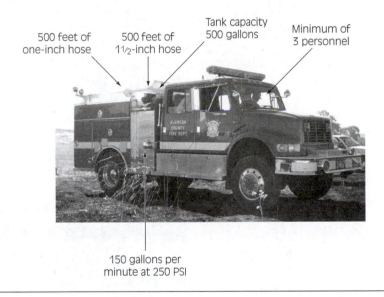

Figure E-10E   Type-Three engine.

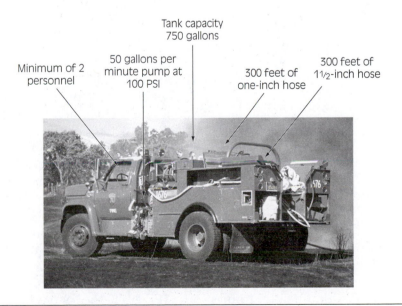

Figure E-10F   Type-Four engine.

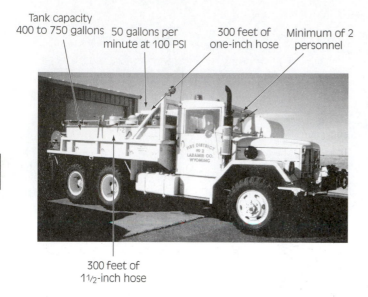

Tank capacity 400 to 750 gallons
50 gallons per minute at 100 PSI
300 feet of one-inch hose
Minimum of 2 personnel
300 feet of 1½-inch hose

Figure E-10G Type-Five engine.

- Type-Six Engine (Figure E-10H)
  Pump rating: 30+ GPM at 100 PSI
  Tank capacity: 150 to 400 gallons
  1.5-inch hose: 300 feet
  1-inch hose: 300 feet
  Personnel: 2 minimum

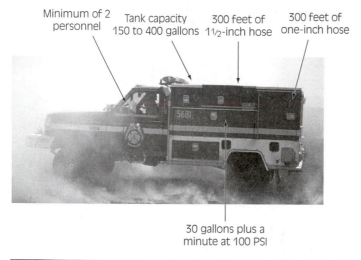

Minimum of 2 personnel
Tank capacity 150 to 400 gallons
300 feet of 1½-inch hose
300 feet of one-inch hose
30 gallons plus a minute at 100 PSI

Figure E-10H Type-Six engine.

- Type-Seven Engine (Figure E-10I)
  Pump rating: 10+ GPM at 100 PSI
  Tank capacity: 50 to 200 gallons
  1-inch hose: 200 feet
  Personnel: 2 minimum

Engine Strike Team Composed of five engines of the same type with common communications and a leader (Figure E-11).

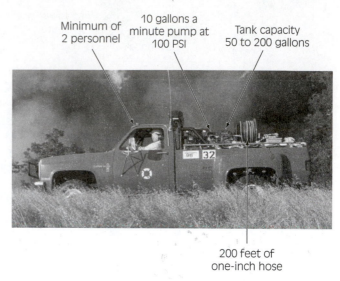

Minimum of 2 personnel
10 gallons a minute pump at 100 PSI
Tank capacity 50 to 200 gallons
200 feet of one-inch hose

Figure E-10I Type-Seven engine.

Figure E-11 Engine strike team.

Extinguishers A portable firefighting appliance designed for use on specific types of fuel and classes of fire. Portable fire extinguishers are designed to fight small fires or unusual fires that are not extinguished with water. Fire extinguishers come in a variety of types and sizes (Figure E-12A). Each type is designed to extinguish certain types of fire. The following is the type of extinguisher for the identified type of fire:

- *Dry Chemical* Ordinary combustibles or Class (A) Fire, flammable liquids or Class (B) fires, and electrical fires or Class (C) fires (Figure E-12B). Shown are the stored-pressure dry chemical parts (Figure E-12C), stored-pressure dry chemical with fixed nozzle (Figure E-12D), dry chemical system

Figure E-12A Various types of extinguishers.

Figure E-12B Stored-pressure dry chemical extinguisher.

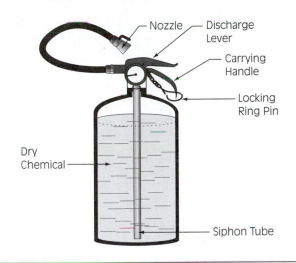

Nozzle — Discharge Lever — Carrying Handle — Locking Ring Pin — Dry Chemical — Siphon Tube

Figure E-12C Parts of a stored-pressure dry chemical extinguisher.

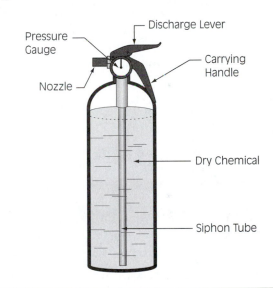

Pressure Gauge — Discharge Lever — Carrying Handle — Nozzle — Dry Chemical — Siphon Tube

Figure E-12D Stored-pressure dry chemical extinguisher with fixed nozzle.

Figure E-12E Dry chemical system with hose reel.

with hose reel (Figure E-12E), a diagram of a dry chemical extinguishing system (Figure E-12F), a cartridge-operated dry chemical extinguisher (Figure E-12G), and a diagram of a cartridge dry chemical extinguisher (Figure E-12H).

E

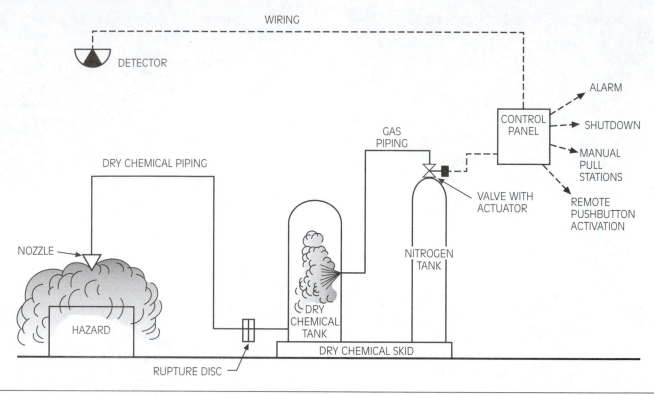

Figure E-12F Dry chemical fire extinguisher system.

Figure E-12G Cartridge-operated dry chemical extinguisher.

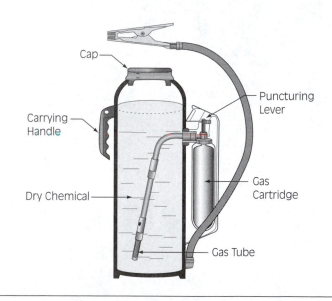

Figure E-12H Parts of a cartridge-operated dry chemical extinguisher.

- *Water* Ordinary combustibles or Class (A) fires. Shown are the stored-pressure water extinguisher (Figure E-12I) and diagram (Figure E-12J), a pump tank water extinguisher (Figure E-12K) and a backpack water extinguisher (Figure E-12L). See also *Backpack Pump* and *Bladder Bag*

E

Figure E-12I   Stored-pressure water extinguisher.

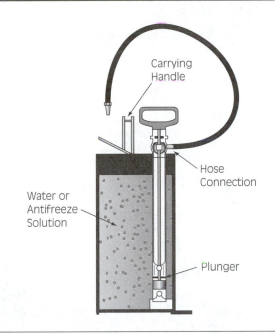

Figure E-12K   Pump tank extinguisher.

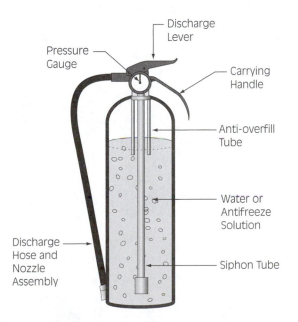

Figure E-12J   Parts of a stored-pressure water extinguisher.

Figure E-12L   Backpack pump tank fire extinguisher.

- *Foam* Ordinary combustibles or Class (A) fires and flammable liquids or Class (B) fires. Shown is the stored-pressure foam extinguisher (Figure E-12M).

Figure E-12M Stored-pressure foam extinguisher.

- *Carbon Dioxide (CO₂)* Flammable liquids or Class (B) fires and electrical Class (C) fires. Shown is the carbon dioxide extinguisher (Figure E-12N).

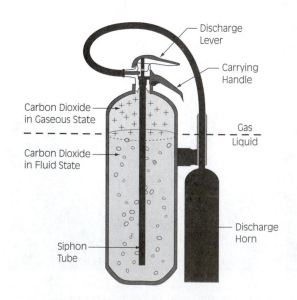

Figure E-12N Carbon dioxide extinguisher.

- *Class K Wet Chemical* Cooking oil or Class (K) fires (Figure E-12O).

Figure E-12O Class (K) fire extinguisher equipment.

- *Dual Agent System* A fire suppression system that uses a combination of foam and dry chemical agents. The combination affords fast knockdown and extinguishment, and it prevents reignition. (Figure E-12P).

Figure E-12P Trailer-mounted dual agent extinguisher system.

- There is no one extinguisher that will extinguish all Class (D) metal fires.

Facepiece The portion of a self-contained breathing apparatus (SCBA) that fits over the face by means of an adjustable harness. It may or may not have the regulator attached to it (Figure F-1). See also *Air Mask*

Figure F-1 Facepiece.

Festoons Cordage and carabiner rope rides that lift the weight of a control line up onto the trackline (Figure F-2). Festoons are best made with parachute or 3-millimeter utility cord and lightweight carabiners. Also called Line Hangers.

Figure F-2 Festoon.

Field Operations Guide (FOG) A guide for the application of the incident command system (ICS) to any planned or unplanned event (Figure F-3).

Figure F-3 Field Operations Guide (FOG).

File A tool with roughened surfaces used to remove surface material from a solid, such as metal or wood. Firefighters use files to sharpen and repair tools, such as chain saws, shovels, mcleods, axes, and so on. There are many types of files used by the firefighter, classified as single- or double-file cuts. Saw files and files intended for use on soft metal are generally single cut. The cut a file makes is referred to as rough, middle or coarse; bastard (regular); second cut (fine); smooth; and dead smooth (super fine or finished).

F

When a file is not cut on one or both sides, these sides are said to be "safe." Also called a "bellied" file. General purpose files have round or square ends, such as a flat file (Figure F-4A) that has a rectangular cross section and is 2 to 20 inches long. Flat files come in regular, blunt, hand or potance, and tapered styles. The half-round file is another common type that comes either in a variety of tapers or with a blunt end. Following are several types of files used by firefighters today:

- **Cant Saw** A single-cut file used for sharpening back saws with M-shaped teeth. Also used for filing an inner angle such as that of a square bolt hole or an open-end wrench. Cant saws are 4 to 12 inches long (Figure F-4B). Also called a Cant File or Lightning File.

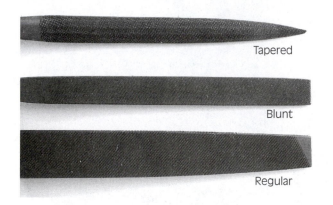

Tapered

Blunt

Regular

Figure F-4A Flat files.

Figure F-4B Cant saw, used for sharpening saws whose teeth are set at less than a 60° angle.

- **Chainsaw** Used for saw chains and to clean and smooth out the gullets of some saws (Figure F-4C).

Figure F-4C Chainsaw file.

- **Crosscut** The rounded back is used to deepen the rounded gullets of saw teeth. The sides are used for filing the teeth (Figure F-4D).

Figure F-4D Crosscut file, for sharpening saws.

- **Flat Chainsaw** Used for lowering and adjusting the depth gauge of saw chains (Figure F-4E). Also called a Raker.

Figure F-4E Flat chainsaw file (raker).

- **Taper** A file that is generally single-cut, 2 to 20 inches long, with a triangular cross section. It is used primarily for sharpening hand saws. The edges of a taper file are slightly rounded to avoid cutting a sharp nick at the base of the saw tooth (Figure F-4F). Also called a Saw File.

Figure F-4F Taper file, for filing saws with 60°-angle teeth.

Fill Station An enclosed area where breathing cylinders are filled. They are designed to protect the operator and bystanders in the event a cylinder ruptures while filling (Figure F-5).

Fire Alarm Control Panel (FACP) A control system for receiving fire alarm signals and initiating action to highlight conditions (alarms and beacons) or institute action to automatically activate fire protection systems.

Fire Blocking Gel A superabsorbent polymer used as a firefighting agent. The fire gels contain water-filled bubbles instead of the air-filled bubbles that we find in Class A foams. These water-filled bubbles allow the product to resist higher temperatures than

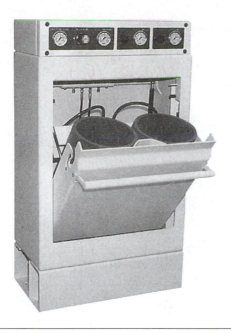

Figure F-5 Fill station.

Class A foams. The gel also seems to have better adhesion qualities. Ground fuels that are coated with gels resist fire for up to 24 hours (Figure F-6).

Figure F-6 Mixing barricade fire gel for aerial delivery to a fire.

Fire Broom A 53.75-inch broom with a hardwood handle and stiff-textured palmyra fiber, used for small grass and underbrush fires (Figure F-7).

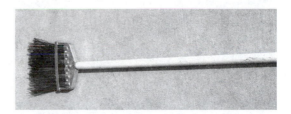

Figure F-7 Fire broom.

Fire Curtain A fire-resistant curtain or screen that can be rapidly lowered to form a barrier from the heat for protection (Figures F-8A–E).

Figure F-8A Fire enclosure for crew modules, open cab engine.

Figure F-8B Fire enclosure—rear view.

Figure F-8C Fire curtain inside cab—outside view.

Figure F-8D Fire curtain—inside cab kit.

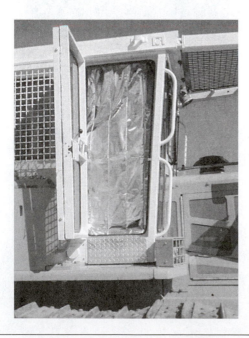

Figure F-8E Fire curtain for tractors, bulldozers, and ATVs to protect operators from heat and fire.

Fire Department Connection Fittings connected to the fire protection system used by the fire department to boost the pressure and/or add water to the system (Figure F-9).

Figure F-9 Fire department connection and post indicator valve.

Fire Hazard Identification A labeling system established by the National Fire Protection Association (NFPA) called the NFPA 704 system, developed to provide a readily identifiable means to ascertain material hazard. The system identifies fire hazards in three main areas: health, flammability, and reaction or instability. The relative ranking in severity of each hazard category is indicated with a numeric value from zero to four (no risk to severe risk). Generally, it consists of a diamond-shaped placard, divided into four smaller diamonds or quadrants. Each quadrant is color-coded and specifically arranged for the three main hazard areas (Figure F-10).

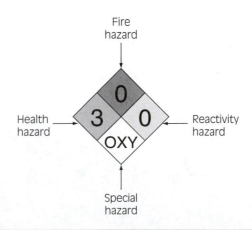

Figure F-10 Fire hazard identification NFPA-704 system.

Fire Hose Float A water-rescue flotation device made from inflated fire hose.

Fire Pack An individual unit of fire tools, equipment, and supplies prepared in advance for carrying on the back (Figure F-11).

Figure F-11 Fire pack.

Fire Shelter An aluminum tent that reflects radiant heat and provides a volume of breathable air in a fire entrapment situation (Figures F-12A–D). Fire shelters should only be used in life-threatening situations.

Figure F-12A Fire shelter.

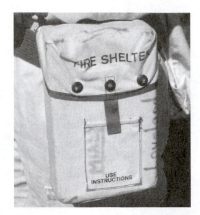

Figure F-12B Fire shelter case.

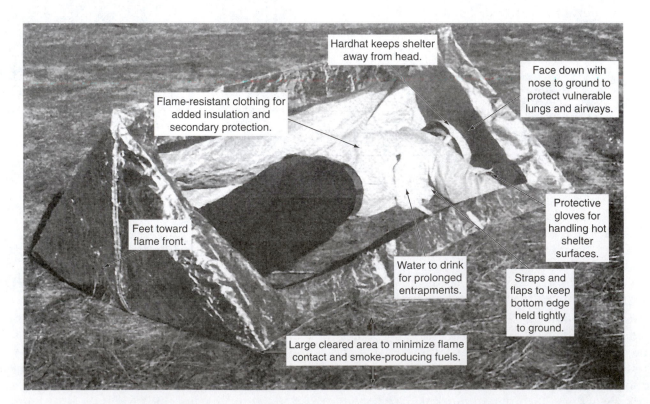

Hardhat keeps shelter away from head.

Face down with nose to ground to protect vulnerable lungs and airways.

Flame-resistant clothing for added insulation and secondary protection.

Feet toward flame front.

Water to drink for prolonged entrapments.

Protective gloves for handling hot shelter surfaces.

Straps and flaps to keep bottom edge held tightly to ground.

Large cleared area to minimize flame contact and smoke-producing fuels.

Figure F-12C Fire shelter that has been deployed.

F

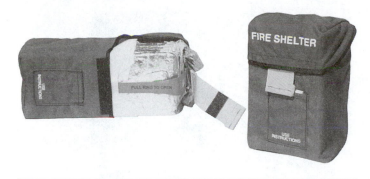

Figure F-12D Fire shelter-new generation.

Fire Shovel *See Shovel*

Fire Station The building in which fire suppression forces are housed (Figure F-13).

Figure F-13 Fire station.

Fire Stream Calculator A handheld sliding scale that shows nozzle pressure calculations on one side and friction loss on the other side (Figure F-14).

Figure F-14 Fire stream calculator.

Firefighter One who suppresses fire (Figure F-15A). Over the years, firefighters have developed many skills and positions and are required to take on many tasks. Fighting fire is just one of the jobs these individuals do today, as shown in Figures F-15B–I.

Fireline Handbook This is a book about the size of a standard paperback that is often carried by officers and crew leaders. It includes charts used to calculate fire behavior, including rate of spread, probability of ignition, flame length, and so on. Most of these charts require a weather kit to obtain current data needed for these calculations. In addition to this there are

Figure F-15A Firefighters.

Figure F-15B Confined space rescue firefighter.

charts with estimations of work potential for various resources, safety reminders, and other useful information including first aid, compass use, hand signals, and more (Figure F-16).

Figure F-15C   Crash fire rescue firefighter.

Figure F-15E   Ice rescue firefighter.

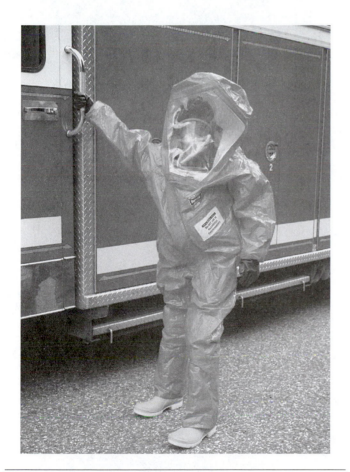

Figure F-15D   HAZMAT firefighter.

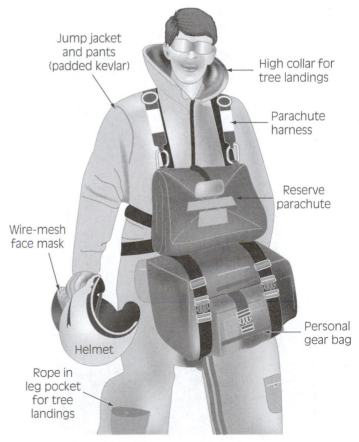

Jump jacket
and pants
(padded kevlar)

High collar for
tree landings

Parachute
harness

Reserve
parachute

Wire-mesh
face mask

Personal
gear bag

Helmet

Rope in
leg pocket
for tree
landings

Firemaxx   A multipurpose tool that has 14 different
built-in tools, such as an axe head, hammer, pry bar,
and spanner wrench within (Figure F-17).

Figure F-15F   Smoke jumper.

F

Figure F-15G Structure firefighter.

Figure F-15I Wildland firefighter.

Figure F-15H Water rescue firefighter.

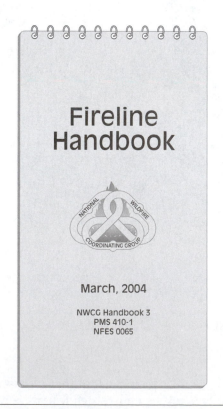

Fireline
Handbook

March, 2004

NWCG Handbook 3
PMS 410-1
NFES 0065

Figure F-16 Fireline Handbook.

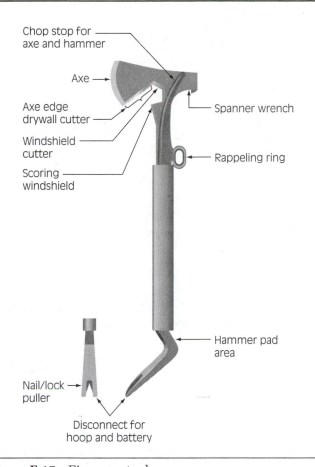

Figure F-17 Firemaxx tool.

Fixed Tank A device mounted inside or directly underneath an aircraft that contains water or retardant for dropping onto a fire (Figure F-18).

Figure F-18 Fixed tank.

Flagging Multicolored ribbon to identify hazards on the fire ground. The Fireline Handbook identifies safety flagging standards. Yellow and black striped ribbon denotes hazards. Hot pink ribbon marked "escape route" in black lettering denotes safety zones and escape routes. Firefighters should check with state and agency policy to verify flagging standards and interagency agreements (Figure F-19).

Figure F-19 Flagging.

Flail Fire suppression tool, sometimes improvised, used in direct attack for smothering flames along a fire edge. It may consist of a green pine bough or wet sacking, or it may be a manufactured tool such as a flap or belting fabric fastened to a long handle (Figure F-20). Also called a Fire Swatter or Flapper.

Flame Thrower Device for throwing a stream of flaming liquid, used to facilitate rapid ignition during burn-out operations. Also known as a Terra Torch.

Flare A colored flare originally designed as a railway warning device and widely used by firefighters

F

Figure F-20 Flail.

Figure F-21C Flare—cap-cover type.

to ignite suppression and prescription fires (Figure F-21A). Also called a Fusee.

Figure F-21A Flare or fusee.

Types of Flares include the pull-tab type (Figure F-21B) and the cap-cover type (Figure F-21C).

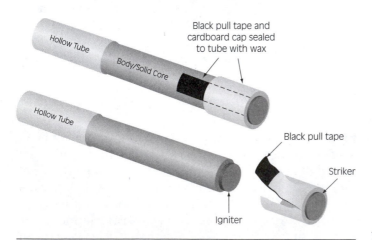

Figure F-21B Flare—pull-tab type.

Flare Launcher This device is made to launch flares during firing operations. It is generally powered by compressed air and has a range of approximately

100 meters. Often these are "home built," using a section of pipe, an air line, and an air compressor (Figure F-22).

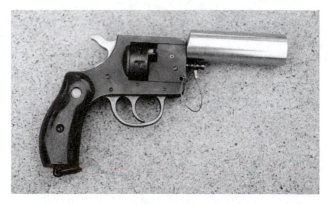

Figure F-22 Flare launcher.

Flex-Beam Dynamometer A device that measures the approximate tension in a rope system (Figure F-23).

Figure F-23 Flex-beam dynamometer.

Float Dock Strainer A floating device attached to a strainer that keeps the suction strainer off the bottom of a fill site (Figure F-24).

Figure F-24 Float dock strainer.

Floor Runner A salvage tool usually made of canvas or similar fabric placed on a floor before firefighters walk on it to protect the floor surface or covering (Figure F-25).

Figure F-25 Floor runner.

Flow Meter A device that eliminates the need for friction loss calculations during pump operations. The flow meter measures the flow of water in gallons per minute (GPM) (Figures F-26A and F-26B).

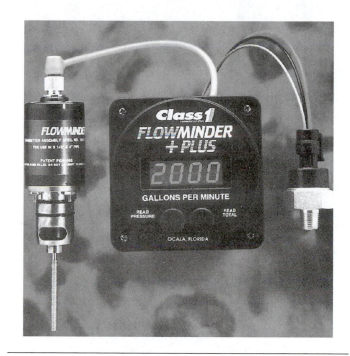

Figure F-26A Flow meter.

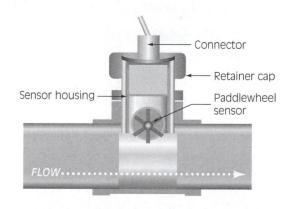

Figure F-26B Flow meter diagram.

Foam An extinguishing agent formed by mixing a foam-producing compound with water and aerating the solution for expansion (Figure F-27A). Foam is applied to any material that is on fire or could potentially catch fire, creating a barrier between the material and the heat and preventing ignition of flammable gases. Foam is commonly used on Class B flammable liquid fires (gas or oil) (Figure F-27B) but is also used in some areas for Class A automobile, wildland, and structure fire applications (Figure F-27C). Types of foam include protein, synthetic,

Figure F-27A Foam.

Figure F-27B Foam applied on Class B fuel fire.

Figure F-27C Application of foam on a Class A fire.

aqueous film forming, high expansion, or alcohol type, as described here:

- **Alcohol Resistant/Aqueous Film Forming Foam (AR/AFFF)** A concentrate suitable for both hydrocarbon and polar-solvent exposures and fuels. The concentrate is a combination of synthetic stabilizers, fluorocarbons, and proprietary additives (Figure F-27D).

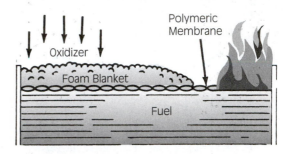

Figure F-27D Alcohol-type foam.

- **Aqueous Film-Forming Foam (AFFF)** A synthetic foam that as it breaks down forms an aqueous layer or film over flammable liquid (Figure F-27E).

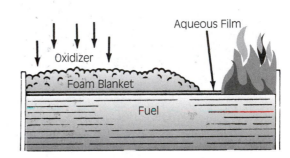

Figure F-27E Aqueous film-forming foam.

- **Fluoroprotein (FP)** Is a protein hydrolysate base combined with fluorochemical surfactants.
- **Fluoroprotein Film-Forming Foam (FFFP)** Combines protein with the film-forming fluorinated surfactants of AFFF to improve on the qualities of both types of foam.
- **High Expansion Foam** Produced by running the foam solution over specially designed netting while forcing air through the netting with a powerful fan called a foam generator high expansion foam blower (Figure F-27F). This creates a foam with an expansion ratio as high as 1,000 to 1.

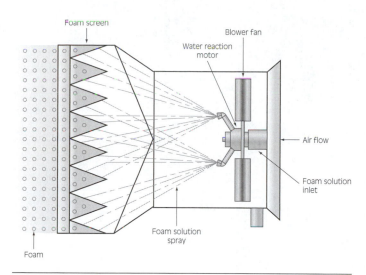

Figure F-27F Foam generator or high-expansion foam blower.

- **Protein Foam** Is a hydrolysate base of hoof and horn, fish scales, etc., combined with stabilizers.

Foam Backpack See *Backpack*

Foam-Balanced Pressure System One type of balanced pressure system uses a pressure vessel with an internal bladder that contains foam concentrate. Water enters the container and exerts pressure on the bladder, expelling the foam concentrate at pressure equal to the ratio controller (proportioner) for discharge. The second type of balanced pressure system uses a separate foam concentrate pump that pumps concentrate at a pressure equal to the pump pressure to a metering valve, which then directs it to a ratio controller (proportioner) connected to the discharge (Figure F-27G).

Foam Blanket A covering of foam applied over a burning surface to produce a smothering effect. It can also be used on non-burning surfaces to prevent ignition (Figure F-27H).

Foam Bucket A container of foam.

Foam Cannon A large foam application nozzle that cannot be handheld due to the size and reaction forces (Figure F-27I).

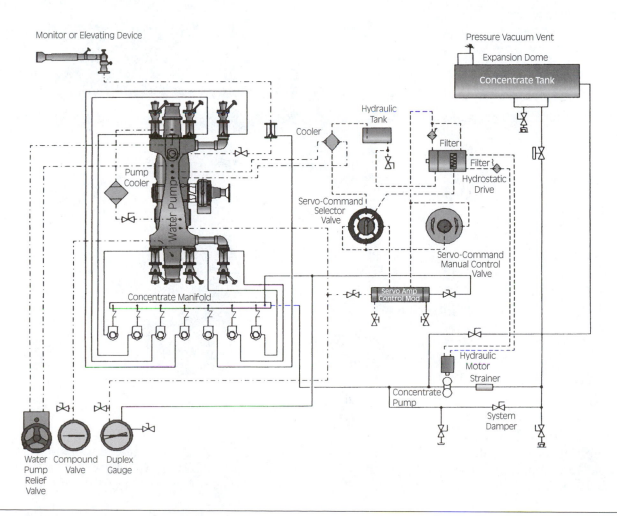

Figure F-27G Foam-balanced pressure system.

F

Figure F-27H   Foam blanket.

Figure F-27I   Foam cannon.

Figure F-27J   Foam manifold on oil tank.

Foam Concentrate   The agent as received from the manufacturer that, when added to water, creates a foam solution.

Foam Eductor   See *Eductor*

Foam Generator   A device that forces air into a foam solution, creating a mass of bubbles. See also *High-Expansion Foam*

Foam Manifold   Allows fire departments to pump foam onto the old-style cone roof of an oil tank and vapor space above the contents using a subsurface foam injection plumbing system (Figure F-27J).

Foam Nozzle   See *Nozzle*

Foam Wrench   A tool that loosens and tightens 70-millimeter foam container caps (Figure F-27K).

Foaming Agent   An additive that reduces the surface tension of water (producing wet water), causing it to spread and penetrate more effectively and which produces foam through mechanical means.

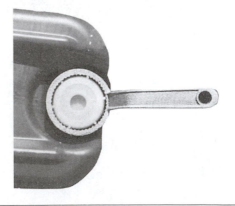

Figure F-27K   Universal foam wrench.

Folding Tank   A portable, collapsible water tank with a tubular frame. Capacity ranges from 500 to 1,500 gallons (Figure F-28).

Figure F-28 Folding tank.

Friction Loss Chart Used to calculate friction loss in hose and appliances (Figure F-29).

Front Mount Pump See *Pump*

Full Protective Clothing See *Personnel Protective Equipment*

HOSE FRICTION LOSS (PSI PER 100 FT)									
G.P.M.	1-1/2"	1-3/4"	2"	2-1/2"	G.P.M.	1-3/4"	2"	2-1/2"	3"
60	8.6	4.9	2.2	.72	200	54.0	24.0	8.0	3.2
70	11.8	6.6	2.9	.98	220	65.3	29.0	9.7	3.9
80	15.4	8.6	3.8	1.3	240	77.8	34.6	11.5	4.6
90	19.4	10.9	4.9	1.6	250	84.4	37.5	12.5	5.0
100	24.0	13.5	6.0	2.0	280	105.8	47.0	15.7	6.3
110	29.0	16.3	7.3	2.4	300		54.0	18.0	7.2
120	34.6	19.4	8.6	2.9	320		61.4	20.5	8.2
130	40.6	22.8	10.2	3.4	340		69.4	23.1	9.2
140	47.0	26.5	11.8	3.9	350		77.8	24.5	9.8
150	54.0	30.4	13.5	4.5	380		86.6	28.9	11.6
160	61.4	34.6	15.4	5.1	400		96.0	32.0	12.8
180	77.8	43.7	19.4	6.5	420			35.3	14.1

DATA SHOWN IS THEORETICAL, ACTUAL LOSS MAY VARY SLIGHTLY BASED ON CONDITIONS

Figure F-29 Friction loss chart.

F

Gas Indicator   Flammable gas indicators (FGIs) are used to measure the gas for which they are calibrated, such as methane (natural gas) using pentane gas (Figure G-1). Also called Combustible Gas Indicators or Sensors.

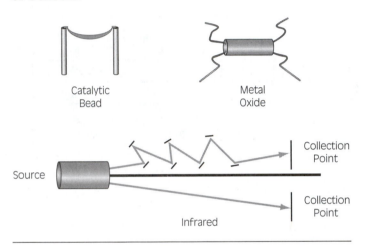

Figure G-1   Gas indicator.

Gas Shut-Off   Usually a slot or part of a tool used to turn gas valves, found at the end of poles or as part of a spanner wrench (Figure G-2).

Figure G-2   Gas shut-off.

Gauge   An instrument that shows the operating conditions of an appliance or piece of equipment. Gauge pressure is measured in pounds per square inch (PSI).

- **Bourdon Gauge**   The type of gauge found on most fire apparatus in which pressure in a curved tube moves an indicating needle (Figure G-3A).

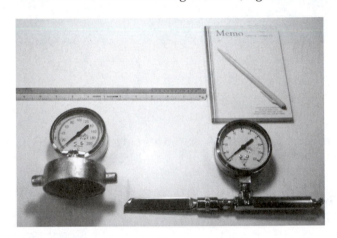

Figure G-3A   Bourdon gauge with a pitot gauge and a cap gauge.

- **Calibrated In-Line Gauge**   Used to test static pressure or flowing pressure in nozzles or hoses (Figure G-3B).

Figure G-3B   Calibrated in-line gauge.

- **Cap Gauge**   Used to determine static or pressure on hydrants, engines (pumpers), or hoses (Figure G-3C). Also called a Static Pressure Gauge.
- **Hydrant Gauge**   A velocity gauge that shows what pressure and how much water will flow from a hydrant (Figure G-3D).

Figure G-3C Calibrated cap gauge.

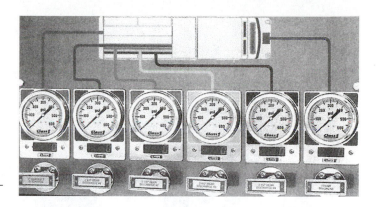

Figure G-3F Pressure gauges.

Figure G-3D Hydrant gauge.

• **Pitot Gauge** A device with an opening in its blade-shaped section that allows water to flow to a bourdon gauge that registers the flowing discharge pressure of an orifice (Figure G-3E).

Figure G-3E Hydrant testing with pitot gauge.

• **Pressure Gauge** Measures pressure without atmospheric pressure. Normally fire department gauges do not measure atmospheric pressure (Figure G-3F).

• **Vacuum Gauge** Shows negative pressure, which is less than atmospheric, usually read in inches of mercury (in. Hg or mmHg). Fire apparatus capable of drafting have at least one gauge that measures vacuum pressure (Figure G-3G). Also called a Compound Gauge.

Figure G-3G Vacuum gauge.

General Order A standing order, usually written, that is communicated through channels to all units and remains in effect until further notice.

Generator Produces electricity for firefighters at the scene of an incident to operate electrical powered tools and lights. Generators come in several different sizes and levels of mobility. Portable generators are powered by gasoline, diesel, or hydraulic motors and generally have 110- and/or 220-volt capacities (Figure G-4).

Figure G-4 Generator.

Gin Pole Pole held upright at an angle by guy lines and used with a block and tackle for lifting (Figure G-5).

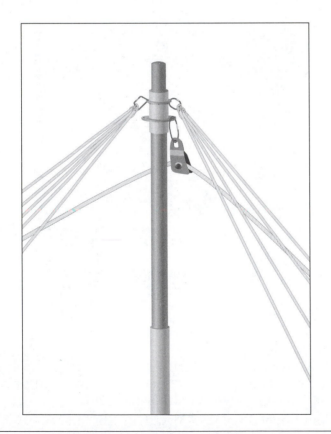

Figure G-5 Gin pole.

Gizmo See *Hose Appliance*

Glas-Master A rescue tool that removes glass from automobiles. It cuts on a pull stroke thereby pulling glass to the outside, away from the victim. It can also be used on fiberglass, wood, and plaster as well as underwater (Figure G-6).

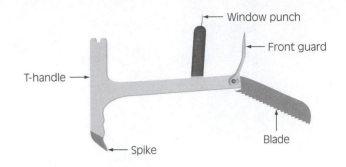

Window punch

Front guard

T-handle

Spike

Blade

Figure G-6 Glas-Master.

Global Positioning System (GPS) A network of 24 navigational satellites (Figure G-7A) orbiting the earth that are used to pinpoint firefighters' position. The satellites orbit at a height of 12,500 miles. A GPS receiver (Figure G-7B) picks up signals from any of these satellites that are above the horizon. It uses information in each signal to determine how far away it is from the satellite. It can calculate its position on the earth's surface when it has information from at least three satellites. A basic GPS receiver shows the latitude and longitude of its position on the screen. A more advanced receiver shows the position on a digital map. Some receivers display extra information, such as the vehicle in which the receiver is installed. The fire service has several uses for GPS—to determine crews' locations, pinpoint spot fires, find landing zones for helicopters, and navigate directly to a requested location for fire retardant or water drops at a wildland fire.

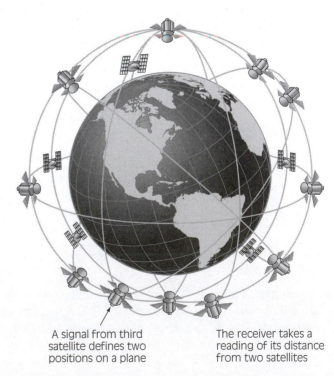

A signal from third satellite defines two positions on a plane

The receiver takes a reading of its distance from two satellites

Figure G-7A GPS satellite positions.

Figure G-7B GPS receiver.

Governor A built-in, pressure-regulating device to control pump discharge pressure by manipulating an engine's revolutions per minute (RPM). It also governs an engine's maximum RPM (Figure G-8).

Figure G-8 Electronic automatic pressure governors.

Grabs Devices constructed to grip rope for raising or holding loads or for climbing. They are made of aluminum, steel, or stainless steel. Many variations have been used over the years, but all have essentially two primary parts—a shell and a cam (Figure G-9).

Figure G-9 Grabs.

Grappling Hook See *Hook*

Gravity Sock Used to capture water from a stream or spring above a fire. The sock is secured by stakes, rocks, or rope at the water source and a hose line is attached. The gathering point should be at least 100 feet above the fire. The sock is 9 inches in diameter and 42 inches long with a 1.5-inch male coupling, and is constructed of heavy-duty, treated canvas duck (Figure G-10).

Figure G-10 Gravity sock.

Gravity Tank An elevated water storage tank for fire protection and community water service. A water level of 100 feet provides a static pressure head of 43.4 PSI minus friction loss in piping when water is flowing (Figure G-11).

Figure G-11 Gravity tank.

Ground Jack A heavy jack attached to the frame or chassis of a ladder truck to provide additional stability before raising the ladder (Figure G-12).

Figure G-12 Ground jack.

Hammer Firefighters use hammers for salvage overhaul, rescue operations, ventilation, and forcible entry. One example is a sledge hammer (Figure H-1).

Figure H-1 Sledge hammer.

Hammer Head Pick A striking tool for breaching concrete and breaking up rubble (Figure H-2).

Figure H-2 Hammer head pick.

Handle Shield A metal cover that protects a tool's wooden handle, such as a flat head axe (Figure H-3).

Figure H-3 Handle shield.

Hardware A type of auxiliary equipment, such as ascent devices, carabiners, descent control devices, pulleys, rings, and snap links (Figure H-4).

Figure H-4 An example of the hardware used by firefighters.

Harness A webbing or rope configuration tied onto people to connect them to a safety tether or other component of a rope. Harnesses are classified in four categories: belt, Class I, Class II, and Class III.

- **Class I Harness** Used for support and stabilization, such as on an aerial ladder or last-chance emergency bailout belt (Figure H-5A).

Figure H-5A Class I harness.

- **Class II** Supports the thighs, waist, and buttocks (Figure H-5B) and is designed for a two-person load, such as when performing a pick-off.

Figure H-5B Class II harness.

- **Class III** Designed to prevent a person from inverting if he or she becomes unconscious (Figure H-5C).
- **Emergency Egress Harness** Made of tubular webbing, specially designed to be put on in less

Figure H-5C Class III harness.

than one minute. See *Chapter E for Emergency Egress Improvised Harness*

- **Hasty Hitch Harness** An improvised webbing harness used when the victim has no reasonable harness, is conscious, and is in good medical condition, and when time is of the essence (Figure H-5D).

The hasty hitch is an improvised harness that can be quickly attached to an ambulatory and conscious patient for lifting and lowering.

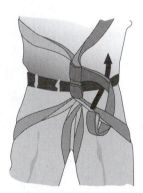

1 The pre-tied loop is placed over the victim's shoulder and between shoulder blades—not around back of neck!

2 Three parts of the webbing are brought forward: (1) over the shoulder, (2) under the arm, and (3) between the legs.

3 All excess webbing is pulled out using the part between the legs. Make sure to pull the other two parts tight against the body. Wrap the long leg section around the other two sections. Finish with a downward wrap of one of the leg loops and a half hitch. This keeps all the connections low and away from the head.

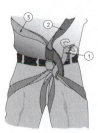

4 Connect a carabiner to three attachment points.

5 Victim can now be carried hanging in the carabiner. **Important:** Keep victim's arms down.

Figure H-5D Hasty hitch.

- **Modified Swiss Seat** An improvised harness that is safer than the emergency egress harness but takes longer to tie. See *Chapter E for Modified Swiss Seat Emergency Egress Harness*

Hatchet A firefighter's hatchet is a smaller version of the Fire Axe or Firefighter's Axe (Figure H-6).

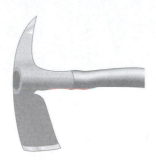

Figure H-6 Hatchet.

Hazardous Material NFPA 704 Placard System A method of identifying hazards at facilities that store hazardous materials (Figure H-7).

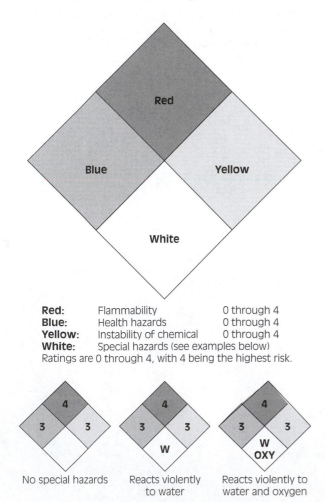

Figure H-7 Hazardous materials NFPA 704 system.

Head Set A radio receiver and transmitter that cover the ears (Figure H-8).

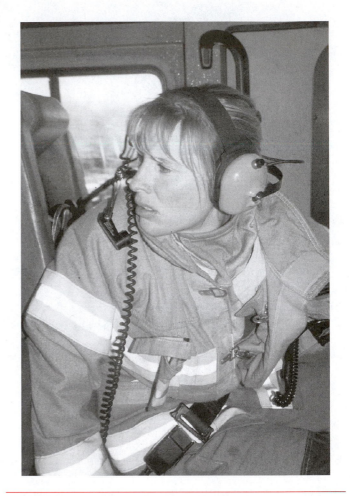

Figure H-8 Head set.

Headlamp See *Light*

Hearing Protection This takes on many forms—foam plugs, rigid earmuffs, and headsets are common examples. Many fire departments prefer the technology that combines hearing protection/intercom/radio microphone into a single headset. See *Combination Head Set*

Heat Detector See *Detector, Temperature*

Heavy Equipment Transport Any ground vehicle capable of transporting a bulldozer, tractor, or other heavy piece of equipment (Figure H-9).

Heel Plate See *Ladder Construction, Heel*

Helibucket Specially designed bucket carried by a helicopter like a sling load and used for aerial delivery of water or fire retardant (Figure H-10).

Helicopter See *Aircraft*

H

Figure H-9 Heavy equipment transport.

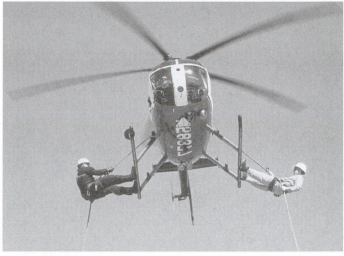

Figure H-11 Helicopter rappelling.

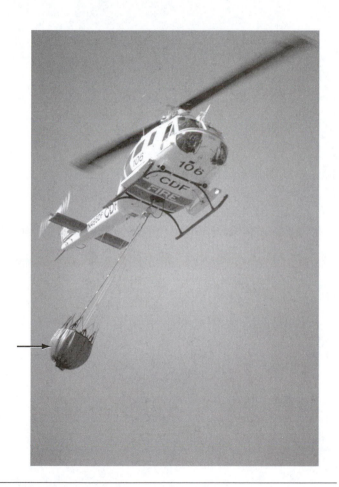

Figure H-12 Helicopter tender .

Helitack Crew Helicopters to transport crews, equipment, and fire retardants or suppressants to the fire line during the initial stages of a fire (Figure H-13).

Figure H-10 Helibucket.

Figure H-13 Helitack crew.

Helicopter Rappelling Technique of landing specifically trained and certified firefighters from hovering helicopters and involves sliding down ropes with the aid of friction-producing devices (Figure H-11).

Helicopter Tender A ground service vehicle capable of supplying fuel and support equipment to helicopters (Figure H-12).

Helitank A specially designed tank, generally made of fabric or metal, fitted closely to the bottom of a helicopter and used for transportation and dropping suppressants or fire retardant (Figure H-14).

Helitanker See *Aircraft*

H

Figure H-14 Helitank under helicopter.

Helitorch See *Aerial Ignition Device*

Helmet See *Personal Protective Equipment*

Helmet Identification Shield The insignia or plaque fastened to the front of a firefighter's helmet that generally displays the name of the department, unit, rank, and number (Figure H-15).

Figure H-15 Helmet identification shield.

Hoisting Device Used to haul victims or rescuers out of a confined space or area (Figure H-16).

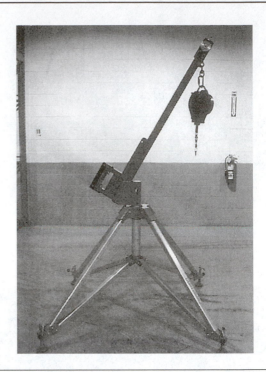

Figure H-16 Hoisting device.

Hook Used in the fire service to pull ceilings or floors and separating other materials.

- **Arson Hook** Used to find hidden fire in deep-seated smoldering fires normally found in dumpsters or garbage dumps. The Arson hook also called a trash and LA hook has two large pointed prongs to grasp debris in large chunks (Figure H-17A).

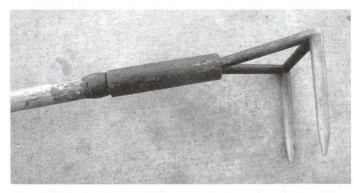

Figure H-17A Arson or trash hook.

- **Boston Rake Hook** Used to pull plaster and lath (Figure H-17B).
- **Chimney Hook** For breaking up ash and creosote build-up in chimneys (Figure H-17C).
- **Clemens Hook** Used for overhaul, salvage, and pulling floor material and wall material (Figure H-17D).

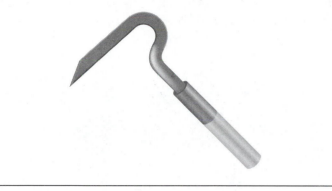

Figure H-17B Boston rake hook.

Figure H-17C Chimney hook.

Figure H-17D Clemens hook.

• **Closet Hook** A small pike pole used in small areas such as in a closet or small room (Figure H-17E).

Figure H-17E Closet hook.

• **Dry Wall Hook** Specifically designed for dry wall removal, but can be used on other wall and ceiling materials such as wood, plaster, and sheet metal (Figure H-17F).

Figure H-17F Dry wall hook.

• **Eckert Hook** Used to open tin ducts, air conditioning systems, and complex metal hoods. It can also be used to pull lath and plaster (Figure H-17G) Also called an EK Hook.

Figure H-17G Eckert or EK hook.

• **Gator Back Hook** Used to pull and cut sheetrock, plaster, and lath (Figure H-17H).

Figure H-17H Gator back hook.

• **Grappling Hook** Attached to a rope for pulling down building materials during overhaul operations (Figure H-17I).

Figure H-17I Four-prong grappling hook.

H

• **Griff Hook**   A shorter version of the New York Hook, sometimes called a Roofman's Hook. (Figure H-17J).

Figure H-17J   Griff hook.

• **Hawk Hook**   Used for removing layered materials, roof materials, and floor panels (Figure H-17K).

Figure H-17K   Hawk hook.

• **Hay Hook**   Used to remove bailed materials or objects like mattresses, beds, and furniture (Figure H-17L).

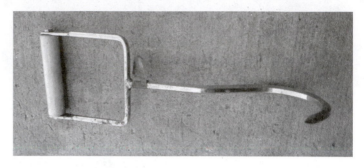

Figure H-17L   Hay hook.

• **Hook and Chain**   Used for pulling, stabilizing, and working with hydraulic rescue tools (Figure H-17M).

Figure H-17M   Hook and chain.

• **Multi-Purpose Hook**   Used for opening up ceilings, floors, walls, moldings, and casings. It can also be used to remove wood, lath plaster, tin, sheet metal, plasterboard, fiber board, and sheetrock (Figure H-17N).

Figure H-17N   Multi-purpose hook.

- **New York Hook** Used for removing molding, baseboards, trim work, and fiberboard (Figure H-17O).

Figure H-17O   New York hook.

- **Pike Pole Hook** This is one of the oldest hooks used in the fire service, used for pulling ceiling and separating building materials (Figure H-17P).

Figure H-17P   Pike pole hook with D-handle.

- **Plaster Hook** Used for pulling lath and plaster (Figure H-17Q).

Figure H-17Q   Plaster hook.

- **Providence Pierce Hook** Used for overhaul and for removing baseboards, door trims, and molding. The pike will penetrate plaster and lath. The hook is curved so it will grab a large amount of material and pull it on the down stroke (Figure H-17R).

Figure H-17R   Providence pierce hook.

- **San Francisco Hook** Used for removing molding, baseboards, trim work, and floorboards (Figure H- 17S).

Figure H-17S   San Francisco hook.

- **Spike Hook** Used for prying in areas difficult to reach during salvage and overhaul operations (Figure H-17T).

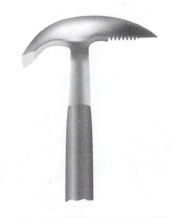

Figure H-17T   Spike hook.

H

H

• **T-Hook** Designed for residential roof work, used for removal of boards and roofing materials and for overhaul operations (Figure H-17U).

Figure **H-17U** T-hook.

• **Trash Hook** See *Hook, Arson*
• **Universal Hook** Used to penetrate several types of building materials (Figure H-17V).

Figure **H-17V** Universal hook.

Halligan Tool See *Bar*

Hose Fire hose is used to deliver water onto a fire and to provide water from hydrants to firefighting apparatus (Figure H-18). The first fire hose was made of leather. Many different materials have replaced leather hose but the basic task—supply and attack—remains the same. Fire hose comes in many different sizes. Small lines such the booster lines and mop up lines are from .75 to 1 inch in diameter. Attack hoses or handline hoses come in 1.5- and 1.75-inch diameters. Medium-diameter hose measures either 2.5 or

Figure **H-18** Hose.

3 inches. Large-diameter hose (LDH) is considered supply hose and comes in 4-, 5-, and 6-inch diameters. Fire hose usually comes in lengths of 25-, 50-, and 100-foot sections.

Fire hose has two components—the hose itself and the couplings that connect the hose sections or appliances used with them. Fire hose is made in three types of construction: wrapped, braided, and woven.

• **Booster Hose** Uses high pressure (typically 800 PSI working pressure,) and has a vertical braid construction overlaid with a premium ozone resistant cover. It can also be constructed of a rubber-coated material and it is usually .75 or 1 inch in diameter (Figure H-18A).

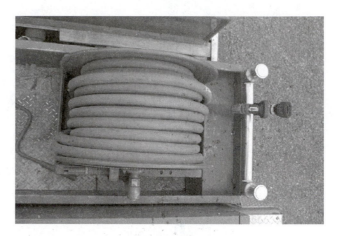

Figure **H-18A** Booster hose.

• **Double-Jacket Hose** Has a second, closely woven jacket for extra durability, safety factor, and higher test pressures (Figure H-18B).

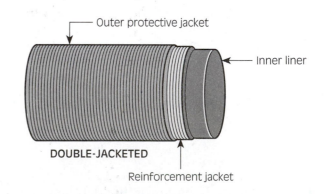

Figure **H-18B** Double-jacket hose.

• **Forestry Hose** Specially designed for use in wildland firefighting. It comes in 1-inch and 1.5-inch sizes (Figure H-18C). Also called Wildland Hose.

Figure H-18C   Forestry or wildland hose.

- **Hard Suction Hose** Manufactured for pump suction service on fire engines. Many different styles are available, from conventional to corrugated to wire-reinforced, and so on. The outer cover is typically made of a long-wearing, ozone-resistant synthetic rubber (Figure H-18D).

Figure H-18D   Hard suction hose.

- **Pin Rack Hose** A thermoplastic-lined hose designed for easy pin rack storage and maintenance. Modern constructions replaced unlined linen hose. This hose is also called Occupant Hose when used in standpipe systems for building occupants. It is usually a 1.5-inch diameter, single-jacket hose (Figure H-18E).

Figure H-18E   Pin rack or occupant hose.

- **Single-Jacket Hose** A thermoplastic or synthetic rubber liner combined with a closely woven, textile jacket. The jacket may be a combination of cotton and polyester or 100% synthetic material (Figure H-18F).

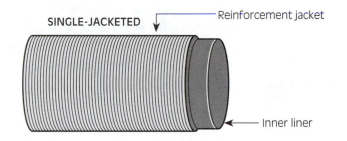

Figure H-18F   Single-jacket hose.

- **Soft Suction Hose** A large-diameter, woven hose used to connect an engine (pumper) to a hydrant. Also known as Soft Sleeve Hose (Figure H-18G).

Figure H-18G   Soft suction or soft sleeve hose.

- **Synthetic Nitrile Rubber Hose** A lightweight, single-jacket hose with a rugged cover, designed for use in industrial environments (Figure H-18H).

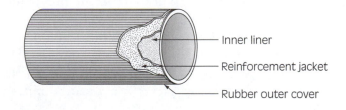

Figure H-18H   Cutaway illustration of a rubber-covered hose.

- **Unlined Fire Hose** Hose commonly of cotton, linen, or synthetic fiber construction without rubber tube or lining, often used for wildfires because of its light weight and self-protecting (weeping) characteristics; such hose is attached to first-aid standpipes in buildings. At a specified flow, friction loss in unlined hose of a stated diameter is about twice that of lined fire hose.

Hose Appliance A device through which water flows including adapters and connectors.

- **Ball Valve** Shaped like a ball to make a smooth transition between off and on (Figure H-19A).

Lug Style Butterfly Valve (Fits between Two Flanges)

Grooved End Butterfly Valve

Figure H-19B Butterfly valves.

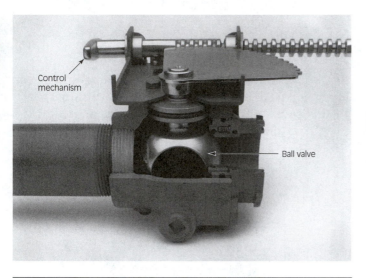

Control mechanism

Ball valve

Figure H-19A Ball valve.

- **Barrel Strainer** Connected to the end of a hard suction hose and shaped like a barrel to keep debris out of the hose line when drafting. See Barrel Strainer in Chapter B for more.

- **Basket Strainer** Connected to the end of a hard suction hose and shaped like a basket to keep debris out of the hose line when drafting. See Basket Strainer in Chapter B for more.

- **Butterfly Valve** Opens and closes in the direction of the handle (Figure H-19B).

- **Drain Valve** Used to relieve pressure from hose connected to the pump or to drain the pump and piping (Figure H-19C).

- **Dump Valve** Used in water tender operations to dump water quickly from the tank (Figure H-19D).

- **Floating Strainer** Used in shallow water to prevent debris being picked up during drafting operations.

- **Four-Way Hydrant Valve** Connected to a hydrant that is directly connected to a supply

Figure H-19C Drain valve.

Figure H-19D Dump valve.

line (Figure H-19E). This valve can be connected to a second engine without interrupting the water supply as the water pressure is boosted. Also called a Hydrant Valve or Switch Valve (Figure H-19F).

Figure H-19E Four-way hydrant valve.

Figure H-19F Switch valve.

- **Gated Valve** Opens and closes using a gate to turn the water supply off and on. This valve is usually controlled by a crank-type handle (Figure H-19G).

Figure H-19G Gated valve.

- **Gated Wyes** Used for connecting one hose line to two hose lines (Figure H-19H).

Figure H-19H Gated wyes.

- **Gizmo** A device inserted into a hose line that restricts water flow when hose lines are laid downhill (Figure H-19I).

Figure H-19I Gizmo.

H

- **Heavy-Duty Appliance** Master stream equipment with large tips generally fed by two or more separate hoses lines (Figure H-19J).

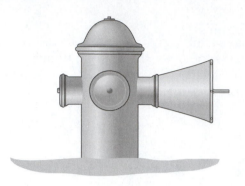

Figure H-19L Hydrant diffuser.

Figure H-19J Heavy-duty appliance.

- **Hose Line T-Valve** Used to add one hose to another or to branch off of an existing hose line (Figure H-19K).

Figure H-19K Hose line T-valve.

- **Hydrant Diffuser** A device connected to a hydrant to diffuse the water coming from the hydrant to prevent damage to the surrounding area (Figure H-19L).

- **Hydrant Valve** See *Four-Way-Hydrant Valve* or *Gated Valve*
- **Hydraulic Ejector** Used to draft water with a vertical lift from 40 to 250 feet, depending on the type of pump and ejector used. The ejector can be used up to several hundred feet from the pump when the pump cannot be spotted within the length of the drafting hose to the water source. See *Chapter E Ejector*
- **Intake Valve** Connects a water supply to a piece of fire apparatus (Figure H-19M).

Figure H-19M Intake valve.

- **Low-Flow Strainer** A strainer designed to perform in low water areas (Figure H-19N).
- **Nozzle** Used to direct a pressurized water stream in a desired pattern, density, and direction. See Nozzle in Chapter N for more.

H

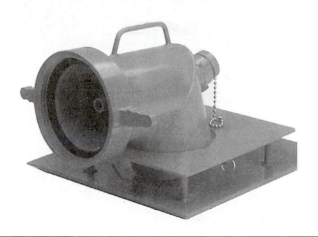

Figure H-19N Low-flow strainer.

- **Relief Valve** A pressure-control device on a pump to prevent excessive pressure by releasing it to a safe discharge location. Relief valves are arranged to automatically bypass or dump water when pressure is exceeded to avoid hazardous conditions. Shown in Figure H-19O is a discharge relief valve and an intake relief valve, along with a field adjustable intake relief valve (Figure H-19P), and a pressure relief valve (Figure H-19Q).
- **Siamese** Used to connect hose lines into one or more hose lines (Figure H-19R).

Figure H-19P Field adjustable intake relief valve.

- **Water Thief** Used for laying, extending, or adding lines without interfering with the operation of other lines (Figure H-19S). Also called a Distribution Manifold when used with large-diameter hose (Figure H-19T).

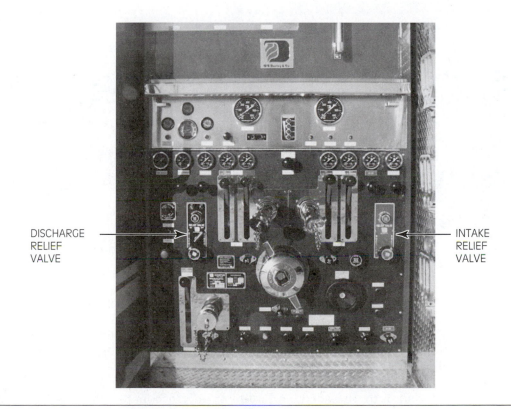

DISCHARGE RELIEF VALVE

INTAKE RELIEF VALVE

Figure H-19O Discharge relief valve and intake relief valve.

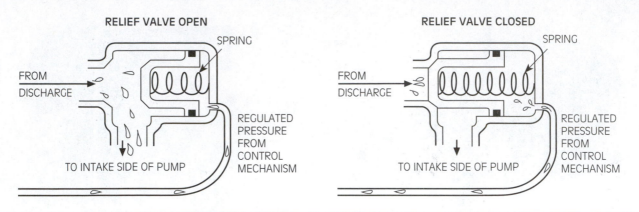

Figure H-19Q Pressure relief valve.

Figure H-19R A variety of Siamese sizes and configurations.

Figure H-19S Water thief.

Figure H-19T Distribution manifold.

Hose Caps See *Hose Fittings*

Hose Coupling Allows hose and appliances to be joined or connected using a set of connection devices. Couplings are divided into threaded and non-threaded types. Threaded couplings use a screw thread that secures the two sections of hose together and come in different thread types such as national hose thread (NH), nation pipe tapered (NPT), and national pipe straight hose (NPSH), while non-threaded couplings use locks and cams. Couplings are made of brass, aluminum, or an alloy called pyrolite or polycarbonate, which is lighter than brass but more resistant to bending. Some couplings have a quick coupling connection, called quarter-turn (Figure H-20A) or storz couplings (Figure H-20B). Threaded couplings have an indicator called a Higbee Indicator or Higbee Cut (Figure H-20C), lined up prior to connecting the hose. Each coupling has a lug, such as a rocker lug (Figure H-20D) or pin-lug coupling (Figure H-20E). Threaded couplings are further separated into two different types: one with external threads (male) and the other with internal threads (female) with a matching thread type (Figure H-20F). Some couplings will have both quick connects and threaded ends.

Figure H-20A   Quarter-turn coupling.

Figure H-20B   Storz quick connect coupling.

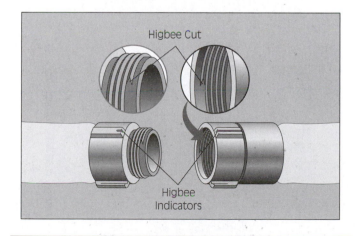

Figure H-20C   Higbee cuts and indicators.

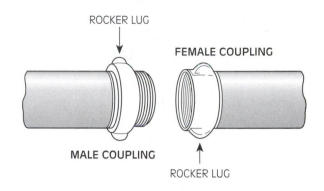

Figure H-20D   Rocker lug.

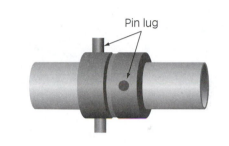

Figure H-20E   Pin lug coupling.

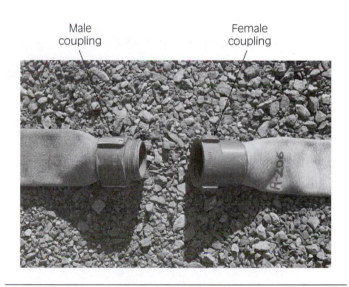

Figure H-20F   Male and female couplings.

Hose Adapter   An adapter for connecting two unequal sizes or different types of threaded ends, caps, or plugs. The following are examples of different types of hose fittings.

- **Cap**   To close off the male end of a hose or outlet (Figure H-21A).
- **Decreaser**   To connect small hoses to large hoses (Figure H-21B).

Figure H-21A Hose cap.

Figure H-21B Two examples of decreasers.

- **Double Female** To connect two male ends together (Figure H-21C).

Figure H-21C Double female.

- **Double Male** To connect two female ends together (Figure H-21D).
- **Plug** To close off the female end of a hose or outlet (Figure H-21E).
- **Reducer** To connect large hose to small hose (Figure H-21F).
- **Universal Storz Coupling** Works with both storz and threaded couplings.

Figure H-21D Double male.

Figure H-21E Hose plug.

Figure H-21F Hose reducer.

Hose Tool An accessory that helps firefighters move or operate hose lines.

- **Hose Belt** A leather belt or nylon strap used for securing and handling charged hose lines or tools or for tying off a ladder. See also *Hose Strap*
- **Hose Bridge** A device that allows vehicles to pass over a section of hose without damaging it (Figure H-22A).

Figure H-22A Hose bridge.

- **Hose Cart** A hand or flat cart modified to carry hose and other equipment around large buildings or in high-rise buildings (Figure H-22B).

Figure H-22B Hose cart.

- **Hose Clamp** Used to control the flow of water by squeezing or clamping the hose shut. Shown in Figures H-22C–E are the Hebert hose clamp, the hose shut-off clamp, and the wildland fire hose clamp (also called the forestry hose clamp).

Figure H-22C Hebert hose clamp.

Figure H-22D Hose shut-off clamp.

Figure H-22E Wildland fire or forestry hose clamp.

- **Hose Dryer** An enclosed cabinet containing racks on which fire hose can be dried (Figure H-22F).

Figure H-22F Hose dryer.

- **Hose Hoist** A metal frame with a rope attached to fit over a windowsill or edge of a roof with two rollers to allow the hose to roll over the edge, preventing chafe (Figure H-22G).

Figure H-22G Hose hoist.

H

H

- **Hose Jacket** Used to stop leaks when a hose is charged with water. Hose jacket may be a metal or leather device that is fitted over the leaking area and either clamped or strapped together to control the leak (Figure H-22H).

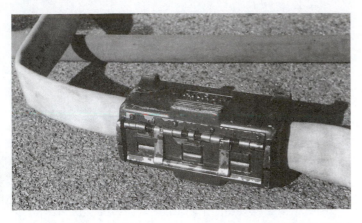

Figure H-22H Hose jacket.

- **Hose Ramp** A device into which a hose sits so vehicles can drive over the hose (Figure H-22I).

Figure H-22I Hose ramp.

- **Hose Reel** A rotating drum used to wind up hose for storing and dispensing (Figure H-22J).

Figure H-22J Hose reel.

- **Hose Roller** A hand-crank device used to roll up wildland or forestry hose (Figure H-22K). It is also the term for a device that squeezes air from the hose (Figure H-22L), as shown in Figure H-22M.

Figure H-22K Wildland hose roller.

Figure H-22L Hose roller for large-diameter hose.

Figure H-22M Firefighters using a hose roller to remove air and water from hose.

- **Hose Rope Tool** A rope spliced into a loop with a large metal hook at one end and a 2-inch ring at the other end. It is used to tie in hose and ladders, for carrying hose, and for many other tasks requiring a short piece of rope (Figure H-22N).

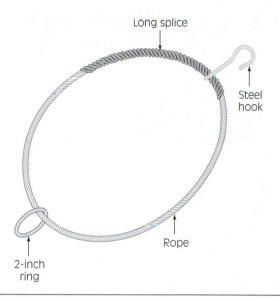

Figure H-22N Hose rope tool.

Figure H-22P Hose strap on ladder.

- **Hose Spanner** Used to tighten or loosen couplings. It is also useful as a pry bar, door chock, gas valve control, and for many other alternative uses. See Spanner in Chapter S for more.

- **Hose Strap** Used for securing hose to ladders (Figures H-22O and P) and in handling charged hose lines. Two hose straps hooked together can be used to carry rolled carpet or other heavy objects.

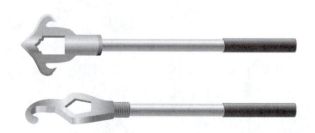

Figure H-22Q Two types of adjustable hydrant wrenches.

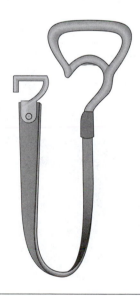

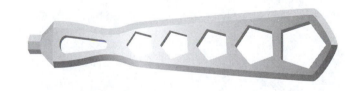

Figure H-22R Five-hole hydrant wrench.

Figure H-22O Hose strap.

- **Hydrant Wrench** A tool used to operate the valves on a hydrant; may also be used as a spanner wrench. Shown are the adjustable hydrant wrench (Figure H-22Q), and the five-hole hydrant wrench (Figure H-22R).

Hose Pack A bundle of hose with tools carried by firefighters for use in high-rise buildings (Figure H-23A), for transport down long driveways and alleys, or at wildland fires (Figure H-23B).

Hose Storage Rack Used to store and dry hose (Figure H-24).

H

Figure H-23A High-rise hose pack.

Figure H-23B Wildland hose pack.

Figure H-24 Hose rack.

Hose Tower A tower or building in which hose is hoisted to let it drain and dry (Figure H-25).

Figure H-25 A firefighter preparing hose to raise up a hose tower.

Hose Washer A device that cleans hose. There are several types of hose washers. Shown are the hydrant hose washer (Figure H-26A) and the automatic hose washer (Figure H-26B).

Figure H-26A Hose washer on a hydrant.

Figure H-26B Automatic hose washer.

Hotshot Crew Intensively trained fire crew used primarily in hand line construction (Type one). See Crew in Chapter C for more.

Hux Bar See *Bars*

Hydrant An upright, pipe-shaped casting with an outlet or spout that is connected to a water supply main to enable the provision of water to fight a fire using hoses and mobile fire apparatus.

Fire hydrants are color coded to indicate the available flow (in gallons per minute (GPM) from the hydrant.

1,500 GPM or greater	Light Blue
1,000 to 1,499 GPM	Green
500 to 999 GPM	Orange
500 GPM or Less	Red

• **Dry Hydrant** A piping system from a drafting source that provides a connection to a static water source that is below grade, such as a lake or pond

(Figure H-27A). In locations subject to freezing conditions, the supply valve and piping are placed below the frost line. Shown in Figures H-27B–D are a dry barrel hydrant, a dry barrel hydrant installation, and a dry barrel hydrant next to a water source.

Figure H-27A Dry hydrant.

Figure H-27B Dry barrel hydrant.

Figure H-27C Dry hydrant installation.

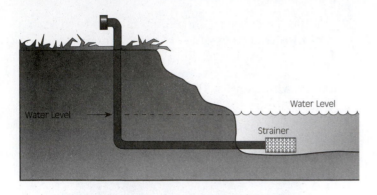

Figure H-27D Dry hydrant next to a water reservoir.

• **Wall Hydrant** A hydrant mounted on the wall of a building after the water line has been run into the building (Figure H-27E).

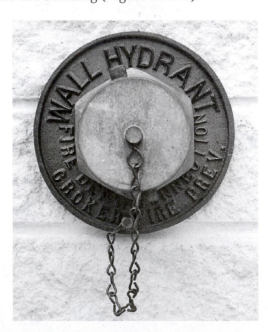

Figure H-27E Wall hydrant.

• **Wet Barrel Hydrant** A hydrant that has water up to the valves of the hydrants, used in areas that are not subject to freezing (Figure H-27F). Shown here is a schematic of a wet barrel hydrant (Figure H-27G).

Figure H-27F Multi-valved wet barrel hydrant.

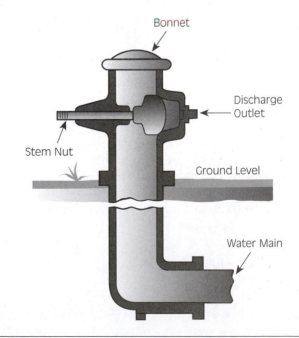

Figure H-27G Schematic of wet barrel hydrant.

• **Yard Hydrant** A fire hydrant installed at industrial facilities, typically with individual valves installed at each hose connection (Figure H-27H).

Figure H-27H Yard hydrant.

Hydrant Finder A pole with reflective red striping tape mounted on a hydrant so the hydrant can be found in heavy snow or weeds (Figure H-28).

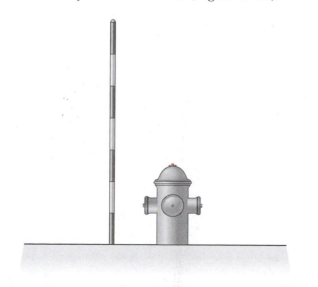

Figure H-28 Hydrant finder.

Hydrant Flag A flag on the end of a pole mounted on a hydrant so the hydrant can be found in high weeds or snow (Figure H-29).

Figure H-29 Hydrant flag.

Hydrant Marker A marker in or near the street that is usually blue and can easily be spotted by firefighters (Figure H-30).

Figure H-30 Hydrant marker.

Hydrant Shut Off A long-handled tool with an attachment that can turn off hydrants from the street below ground (Figure H-31).

Figure H-31 Hydrant street shut-off tools.

Hydraulic Pump Used to provide pressure to hydraulic tools (Figure H-32).

Figure H-32 Hydraulic pumps for extrication equipment.

Hydraulic Spreader A handheld hydraulic tool used for forcible entry, such as power hydraulic spreaders (Figure H-33A), smaller, handheld pump spreaders with remote spreader (Figure H-33B), and the handheld unit with the integral built-in spreader (Figure H-33C), also called a Hydra Ram.

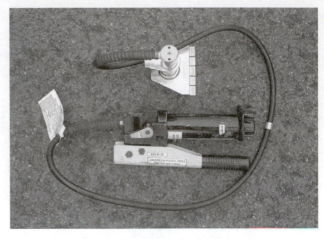

Figure H-33B Handheld pump with remote spreaders.

Figure H-33A Power hydraulic spreaders.

Figure H-33C Handheld hydraulic spreader with integral built-in spreader, also called a hydra ram.

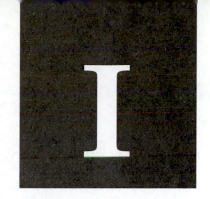

Increaser A device that connects a male outlet to a female inlet of a larger diameter hose with the same thread (Figure I-1).

Figure I-1 Increaser.

In-Line Balanced Proportioning System (ILBP) A foam concentrate proportioner system used with positive displaced foam concentrate pumps and atmospheric foam concentrate storage tanks (Figure I-2).

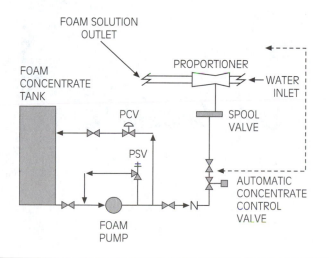

Figure I-2 In-line balance proportioner.

Intake Screen Used to prevent debris from entering a pump.

Ionization Detector See *Detector*

Inflatable Rescue Boat (IRB) See *Boat*

Irons A slang term used to describe a prying tool carried with a striking tool. The most common set of irons is the Halligan tool carried with an axe (Figure I-3).

Figure I-3 Irons.

Jet Pump A valve that draws water out as opposed to just using gravity in a non-assisted system, also called a Jet Siphon (Figure J-1A). It is a device that moves water quickly without generating a lot of pressure (Figure J-1B) from one portable tank to another or to assist in the quick off-loading of water tender water (Figure J-1C).

Figure J-1A Examples of jet pumps.

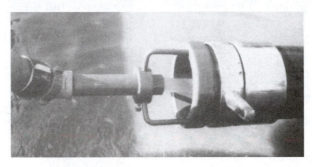

Figure J-1B Jet siphons use very little water and pressure to move large quantities of water.

Figure J-1C Jet siphons move water from one drop tank to another.

Junction Box An electrical box that provides receptacles to a work area (Figure J-2).

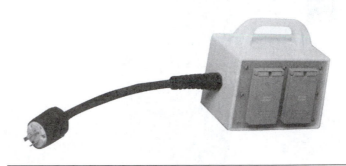

Figure J-2 Junction box.

Jump Line A hose line pre-connected for quick deployment (Figure J-3).

Figure J-3 Jump line.

Key Set A set of keys manufactured to conform with the majority of elevator manufacturer's access keys (Figure K-1).

Figure K-1   Elevator key set.

Key Tool A two-sided, flat or square steel tool used to gain entry by manipulating locks, (Figure K-2). Shown here is a drawing of a key tool in use (Figure K-3).

Figure K-2   Key tools used for forcible entry.

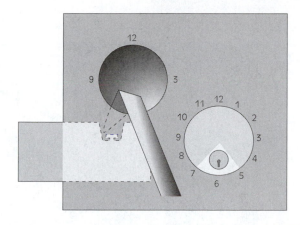

Figure K-3   Key tool being used on a lock.

Knox Box A secured key box that is mounted on the outside of buildings for firefighters to access. The knox or key box contains keys to access the building (Figures K-4A and B).

Figure K-4A   Knox or key box.

Figure K-4B   Firefighters inspecting a knox box.

Kortick Tool A wildland firefighting tool used for fire line construction. It is a combination of a rake and a hoe. See also *McLeod*

The kortick tool is similar to the Mcleod tool but it is heavier and the handle is attached to the blade with a threaded wing nut, which makes this tool much easier to transport.

K-Tool A forcible entry tool designed to pull out lock cylinders and expose the mechanism in order to open the lock with various key tools (Figure K-5).

K-Tool

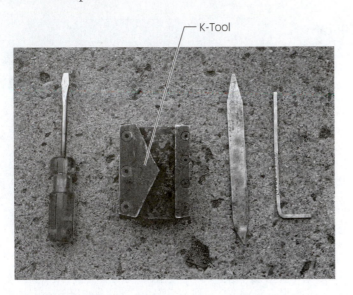

Figure K-5 K-tool shown with key tools.

K

Ladder A structure of two long sides crossed by parallel rungs, used as a means of ascent and descent. The fire service ladder is a tool used for many purposes. The primary purpose is to gain access to elevated locations. The fire service ladder can also be used for shoring (Figure L-1), to construct pools for catching water spills (Figure L-2), to channel water (Figure L-3), to provide elevated streams, to make an A-frame for hoisting (Figure L-4), and for climbing over fences (Figure L-5). There are truck-mounted ladders and ground ladders. Here are examples of truck-mounted ladders:

- **Aerial Ladder** An apparatus-mounted ladder capable of reaching heights of up to 100 feet or more (Figure L-6A). Shown here are the aerial ladder raising mechanisms (Figure L-6B) and an illustration of the aerial ladder position terminology (Figure L-6C).

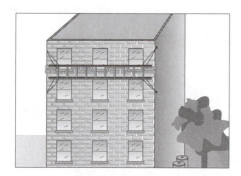

Figure L-1 Ladder used as a shoring. A ladder secured to substantial objects by ropes can assist in stabilizing a structural defect as an emergency measure.

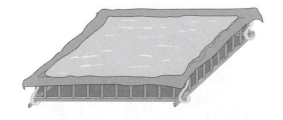

Figure L-2 Ladders being used for an emergency water pool or collection area.

Figure L-3 Ladders with a salvage cover, plastic sheet, or tarpaulin used as a chute to divert and discharge water.

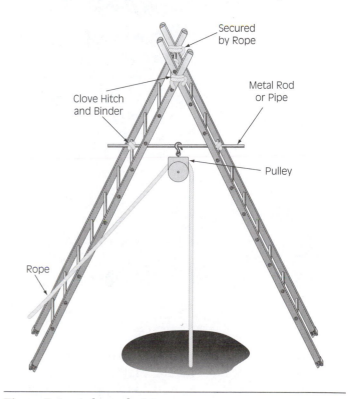

Figure L-4 A-frame hoist.

Figure L-5 Ladder used for climbing over a fence.

Figure L-6A Aerial ladder.

• **Articulating Boom Ladder** An apparatus with a series of booms and a platform on the end, maneuvered into position by adjusting the various boom sections into place.

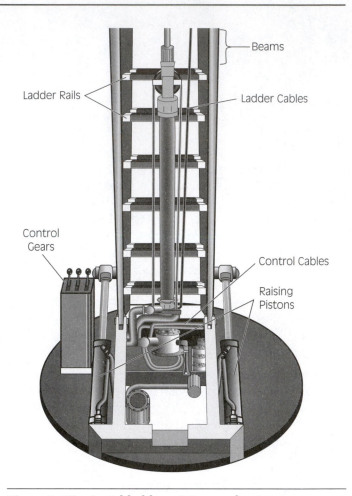

Beams

Ladder Rails

Ladder Cables

Control Gears

Control Cables

Raising Pistons

Figure L-6B Aerial ladder raising mechanisms as seen from under a raised bed.

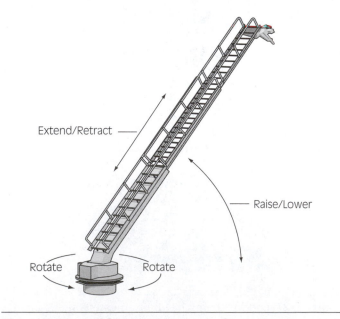

Extend/Retract

Raise/Lower

Rotate   Rotate

Figure L-6C Ladder position terminology.

- **Tower Ladder** A telescopic boom with a mounted basket capable of holding 750 to more than 1,000 pounds (Figure L-6D).

Figure L-6D Tower ladder.

There are also several types of ground ladders.

- **A-Frame Combination Ladder** A combination ladder that can be used in various configurations. When nested, it is easily stored and can also serve as a small extension ladder (Figure L-6E).

Figure L-6E A-frame combination ladder used as a extension ladder.

When articulated into an A shape, it becomes a step ladder. Shown is the combination ladder being converted to a step ladder (Figure L-6F), along with a photograph of the combination ladder as a step ladder (Figure L-6G).

Figure L-6F A frame combination ladder being converted from an extension ladder to an A-frame step ladder.

Figure L-6G A frame combination ladder as an A-frame step ladder.

L

- **Bangor Ladder** Used primarily for elevated access in remote locations where aerial apparatus cannot be placed, Bangor or pole ladders require up to six firefighters and are generally used when no other means of access is available. Ground extension ladders that exceed 40 feet are required to have stay poles, also called tormentor poles. Bangor ladders usually don't exceed 50 feet. The stay poles are used for raising, and, once the ladder is in a raised position (Figure L-6H), they cannot be used to support any weight on the ladder.

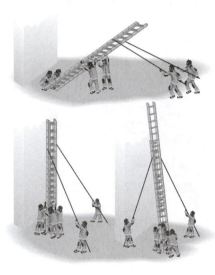

Figure L-6H Bangor or pole ladder.

- **Extension Ladder** Consists of two or more ladders that operate as a unit. The bed ladder acts as the nest for the movable fly ladder. In the two-piece extension ladder, the fly ladder slides along channels built into the bed ladder (Figure L-6I).

2nd Fly Section

Halyard or Cable

1st Fly Section

Bed Ladder

Figure L-6I Typical multifly halyard hoisting pulley arrangement.

- **Folding Ladder** A collapsible ladder (Figures L-6J and K) that can be transported into narrow and confined places. A folding ladder comes in 8-, 10-, and 16-foot lengths. This ladder is also called an attic ladder, suitcase ladder, or closet ladder.

Figure L-6J Folding ladder in the collapsed position.

Figure L-6K Folding ladder in the opened position.

- **Pompier Ladder** This ladder is no longer an approved ladder with NFPA Standard 1932. This ladder is a single beam ladder with rungs extending from the central beam axis (Figure L-6L), also called a scaling ladder.

Figure L-6L Pompier or scaling ladder.

- **Roof or Hook Ladder** A straight or wall ladder with a set of retractable hooks at the tip end. When operating on a sloped roof, it enables a firefighter to work with more secure footing (Figure L-6M). Shown here is a firefighter using a roof ladder (Figure L-6N).

Figure L-6M Hook ladder or roof ladder.

Figure L-6N Firefighter using roof ladder.

- **Straight Ladder** A fixed-length ladder usually between 12 and 20 feet long. It is generally long enough to gain access into first- and second-floor windows. This ladder is also called a wall ladder (Figure L-6O). Shown here is an illustration of straight ladder terminology (Figure L-6P).

L

Figure L-6O Straight wall ladder.

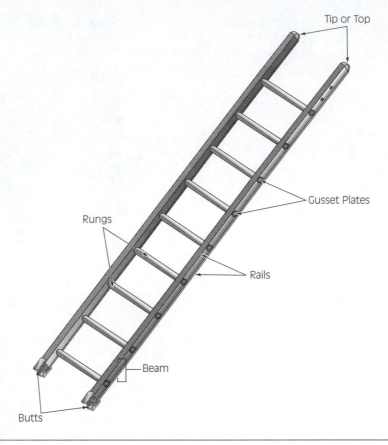

Figure L-6P   Straight ladder terminology.

L

Ladder Belt   See *Belt*

Ladder Company   A combination of a fire truck with an aerial ladder, an assortment of ground ladders and forced entry tools, and the manpower used to staff it, also called a truck company. Ladder trucks can have straight aerial ladders as short as 65 feet or longer ladders with platforms (buckets) on the end. In many departments, ladder companies are responsible for ventilation and forcible-entry duties. A standard ladder company will include an officer, driver/operator, and two firefighters on a ladder truck (Figure L-7).

Ladder Construction   There are many parts to ladders (Figures L-8A–C), although some departments may have different terminology.

• **Beam**   The rail that runs the full length of the ladder from top to bottom, spanned by rungs.

• **Bed Section**   The foundation is usually the part of the ladder that is in touch with the ground or attached to the body of an aerial ladder truck. It is the section from which all other sections are raised in extension ladders. This is also called the bed ladder.

Figure L-7   Ladder company.

• **Dog**   The mechanism in a extension ladder that rides up along with a fly section and engages a rung of the section from which it extends through the use of a spring-loaded lock to prevent retraction. Also called a Pawl, Rung Lock, or Ladder Lock.

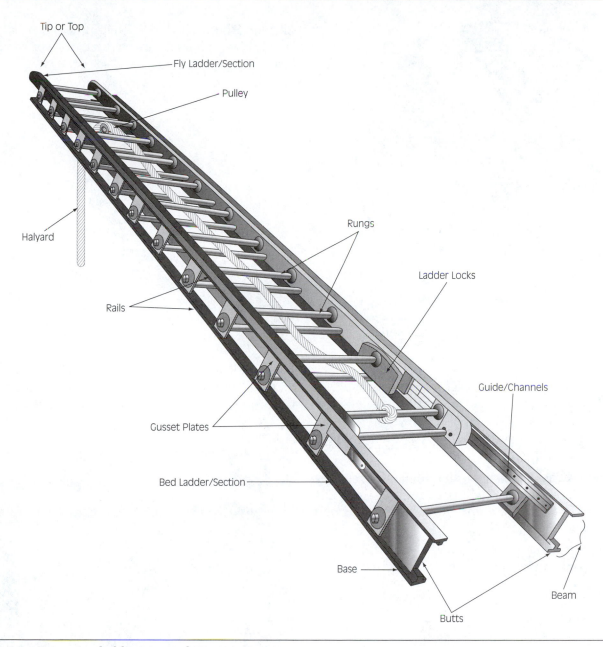

Tip or Top

Fly Ladder/Section

Pulley

Halyard

Rungs

Ladder Locks

Rails

Guide/Channels

Gusset Plates

Bed Ladder/Section

Base

Beam

Butts

Figure L-8A Extension ladder terminology.

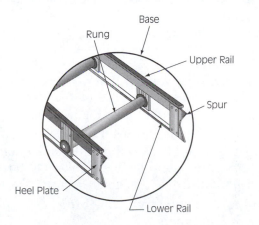

Base

Rung

Upper Rail

Spur

Heel Plate

Lower Rail

Figure L-8B Steel spurs on a ladder of truss construction.

- **Guide** The channel of the bed ladder that permits the fly section to ride up and maintain stability. Also called a Channel.
- **Halyard** The cable or rope, made of nylon, hemp, or steel, used to raise or lower the fly section out of or into the bed section through the use of pulleys.
- **Heal** The bottom of the ladder, usually a reinforced section with points, spurs, or rubber pads to reduce slipping on various surfaces. Also called the Foot, Base, or Butt.
- **Protection Plate** Reinforced metal that is built up at chafing points to avoid weakening created by rubbing and friction wear.

L

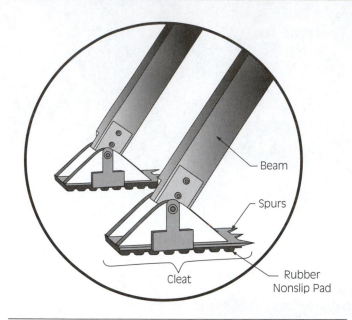

Figure L-8C Swivel shoes with pads and spikes.

- **Pulley** A wheel with a groove through which the halyard passes, used for raising or lowering the fly section of extension ladders.
- **Rail** The upper and lower surfaces of the beams that run lengthwise.
- **Rung** A "step" of a ladder that connects one beam to the other. Generally round, they can also be flattened on the upper side for more secure footing.
- **Sensor Label** A heat-sensitive label affixed to the ladder to alert firefighters that the ladder has been exposed to a potentially damaging heat level and that testing should be performed before it is used again.
- **Spur** One of a pair of pointed shoes that are attached to the base of a ladder to dig into the surface and prevent slippage during use. Also called the Spike, Cleat, Shoe, or Butt Plate.
- **Stop** A limiter built into the bed section to prevent the extension fly section from being overextended. They are found in the shape of solid blocks or angled metal.
- **Tie Rod** Found on wooden ladders, it is a metal rod that secures the two beams and prevents them from spreading apart when the rungs are doweled into the beams. These rods prevent the rungs from pulling out, which would result in a ladder collapse.
- **Tip** The top of the ladder. In an extension ladder, it would be the top of the fly section that attains the greatest height.

Ladder Hook A device that allows a firefighter to attach him or herself to a ladder rung or tower platform (Figure L-9).

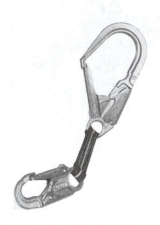

Figure L-9 Ladder hook.

Ladder Pipe A heavy or master stream nozzle attached to the top of an aerial ladder and fed from the ground (Figure L-10). See also *Nozzle*.

Figure L-10 Ladder pipe.

Ladder Truck A vehicle that can carry a full compliment of ground ladders as well as either a telescoping ladder or a platform (Figure L-11).

Lead Plane An aircraft used to make dry runs over the target area to check wing and smoke conditions and topography and to lead air tankers to targets and supervise their drops. See also *Aircraft*

Levels A–D Protective Clothing See *Personal Protective Equipment*

Life Belt See *Belt*

Figure L-11   Ladder truck.

Life Safety Line   See *Rope*

Light   The fire service uses several different types of lights; here are just a few examples:

- **Hand Light**   An individual light that firefighters carry with them (Figure L-12A).

Figure L-12A   Hand light.

- **Head Lamp**   Attached to a helmet to provide hands-free light to a firefighter or rescue worker (Figure L-12B).

Figure L-12B   Head lamp.

- **Light Bar**   An emergency warning light that runs the length of the top of an emergency vehicle (Figure L-12C).

Figure L-12C   Light bar on the top of a command vehicle.

- **Search Light**   A powerful beam that can be mounted on a vehicle or hand held (Figure L-12D). Also called Spot Lights.

Figure L-12D   Search light.

- **Strobe Light**   A light that flashes continuously to indicate a person or vehicle's location (Figure L-12E).

Figure L-12E Strobe light.

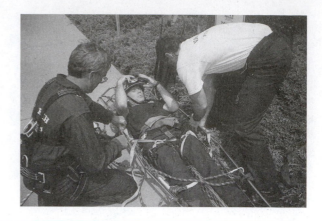

Figure L-13 Litter basket.

- **Telescoping Light** A light that is attached to a generator or fire apparatus that extends up to provide light to a scene (Figure L-12F).

Litter Bridle A device that connects a stokes litter to the raise and lower system (Figure L-14). The two most common litter bridles are the pre-sewn or manufactured and the Yosemite.

Figure L-12F Telescoping lights.

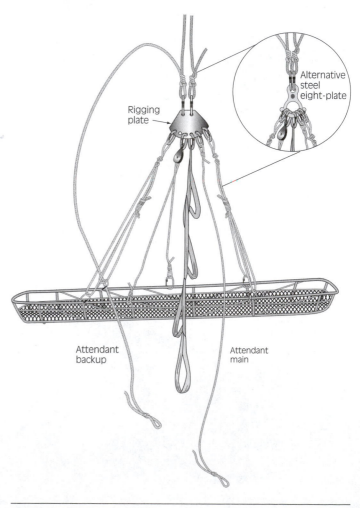

Figure L-14 Litter bridle.

Litter Basket A stretcher made of wire mesh, hard fiberglass, or plastic (Figure L-13). Also called a stokes basket or splint basket.

Load-Sharing Anchor A section of rope or webbing configured to transfer the rope system load to two or more anchors, thereby sharing the load between multiple anchors (Figure L-15).

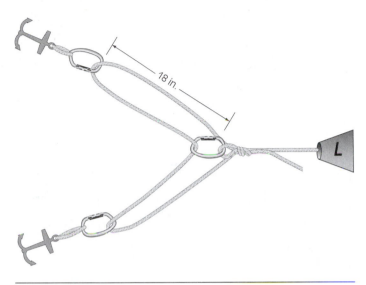

Figure L-15 Load-sharing two-point anchor.

Locking Pliers A tool attached to a chain so a firefighter can hold a lock while it is being cut (Figure L-16).

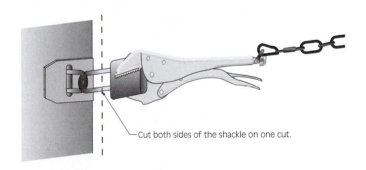
Cut both sides of the shackle on one cut.

Figure L-16 Locking pliers.

Logging Recorder Magnetic tape used by some communications centers to record telephone and radio activity (Figure L-17).

Long Line Line or set of lines, usually in 50-foot increments, used in external load operation that allows a helicopter to place loads in areas in which it could not safely land (Figure L-18).

Figure L-17 Logging recorder.

Figure L-18 Long line being used by a helicopter for bucket drops at a fire.

Male Coupling   See *Hose Coupling*

Mallet   A one-handed striking tool with a large head made of non-marring material such as wood, rubber, or synthetic material. A mallet is generally used for driving another tool or for striking a surface without marring it. Firefighters use mallets for a variety of purposes that require use of a striking tool without causing a spark (Figure M-1).

Figure M-1   Mallet.

Manifold   See *Hose Appliance*

Manual Pull Box   See *Box*

Map   A line drawing, to a scale, of an area of the earth's surface. Unlike photographs, maps are selective and may be prepared to show various quantitative and qualitative facts, including boundaries, physical features, patterns, and distribution. Each point on a map corresponds to a geographical position in accordance with a definite scale and projection. Firefighters use maps for several reasons. Below are descriptions of the different types of maps used in the fire service:

- **Base Camp Map**   Found in the Incident Action Plan and on the bulletin boards at the base camp of a large incident. This map shows the facilities, travel routes, and parking at base camp (Figure M-2A).
- **Contour Map**   The most common method of representing the land's shape and elevation. A contour line is a line of equal elevation on the

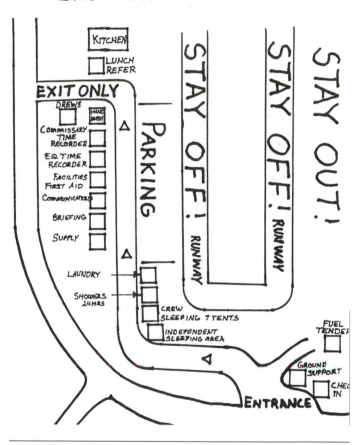

Figure M-2A   Base camp map.

ground that delineates the same elevation above or below a specific reference elevation, usually sea level (Figure M-2B). Figure M-2C shows a map, contour lines representing a saddle.

- **Fire Progressive Map**   Maintained on a large fire to show at any given time the location of the fire, deployment of suppression forces, and progress of suppression (Figure M-2D).
- **Ortho Photo Map**   Is an aerial photograph of the land that depicts terrain and features by color enhancement. The Ortho Photo Map can have contour line overlays.

Figure M-2B Contour map lines.

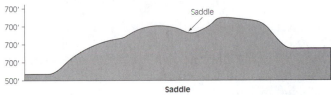

Figure M-2C Contour line representing a saddle.

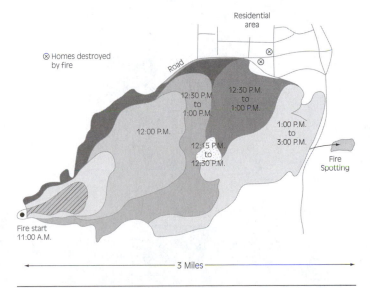

Figure M-2D Fire progressive map.

• **Topographical Map** Shows both the horizontal and vertical (relief) positions of features. Topographical maps are often referred to as quadrangles or quads (Figure M-2E).

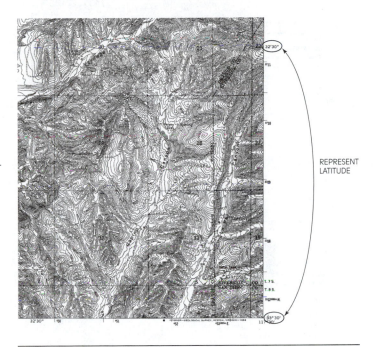

Figure M-2E Topographical map.

Marrying Strap Used to connect the axe with the Halligan tool (Figure M-3) See also *Irons*

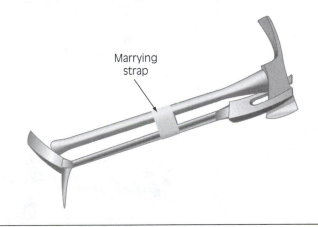

Figure M-3 Marrying strap.

Master Stream A large, fixed stream of water from an appliance, usually mounted on top of fire apparatus.

Material Data Sheet (MSDS) Information sheet for employees that provides specific information about a chemical, with attention to health effects, handling, and emergency procedures (Figure M-4).

M

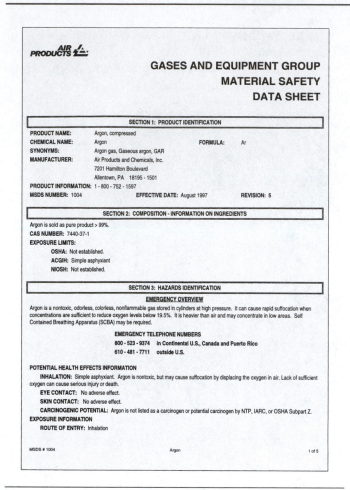

Figure M-4 Material data sheet (MSDS).

Mattock A hand tool for digging and grubbing, with a narrow hoeing surface at one end of the blade and either a pick or cutting blade at the other end. A Mattock tool is used to construct fire lines at wildland fires (Figure M-5).

Figure M-5 Mattock tool.

Maul Firefighters use a maul to drive a wedge in the cut made by a saw in a tree for felling trees. The maul is better designed than an axe for this purpose. Mauls can also be used for rescue and overhaul operations. See *Chapter A Maul Axe*

Maxi Multi-Purpose Tool A wildland firefighting hand tool that can be modified to be a shovel, Pulaski, or pick (Figure M-6).

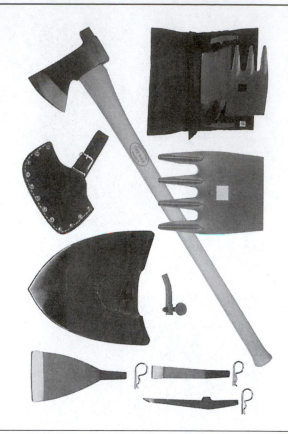

Figure M-6 Maxi multi-purpose tool.

McLeod Tool A scraping and raking tool used by wildland firefighters that has a rake on one side and a scraping head on the other (Figure M-7A). If it has a detachable handle it can also be called a Kortick tool. Also shown is the proper sharpening procedure for the McLeod (Figure M-7B).

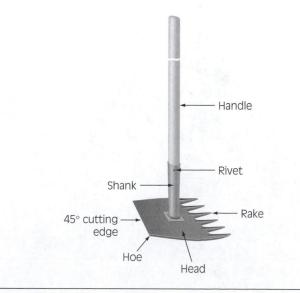

Figure M-7A McLeod tool.

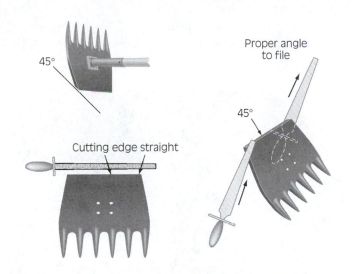

Figure M-7B   McLeod sharpening method.

Megaphone   A voice amplification tool that has a pistol grip used by firefighters for crowd control (Figure M-8).

Figure M-8   Megaphone.

Message Dropper Pouch   A plastic bag designed to hold messages dropped from aircraft. The clear plastic pouch holds ballast sand for weight. It usually has a plastic streamer that is colored international orange or yellow. Wildland firefighters use it to drop information like fire size mapping for firefighters on the ground (Figure M-9).

Figure M-9   Message drop pouch.

Mobile Data Terminal   A communication device that has no information processing capabilities (Figure M-10).

Figure M-10   Mobile data terminal.

Mobile Water Supply Apparatus   A fire service support vehicle, such as a water tender, designed primarily for transporting a specified amount of water to a fire emergency scene (Figure M-11). See also *Water Tender*

Figure M-11   Mobile water supply.

Modified Swiss Seat   See *Harness*

Modular Airborne Firefighting System (MAFFS)   A manufactured unit consisting of five interconnecting tanks, a control pallet, and a nozzle pallet, with a capacity of 3,000 gallons of fire retardant, designed to

M

be rapidly mounted inside an unmodified C-130 (Hercules) cargo aircraft for use in cascading retardant chemicals on wildfires.

Monitor A master stream hose appliance used to deliver large volumes of water at long distances, sometimes mounted on engines and pre-plumbed to the pump (Figure M-12).

Figure M-12 Monitor being used to cool a heated metal surface.

Mop-Up Kit Used during mop-up operations at a wildland fire. The kit usually has enough equipment for three firefighters, including reducers, nozzles, T-valves, applicator wands, and wrenches (Figure M-13).

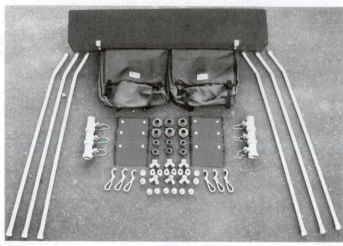

Figure M-13 Mop-up kit.

Municipal Fire Alarm Box See *Box*

Municipal Fire Alarm System A fire alarm system in cities and towns that uses street boxes placed throughout the municipality that transmit alarms to the fire service headquarters monitoring station (Figure M-14).

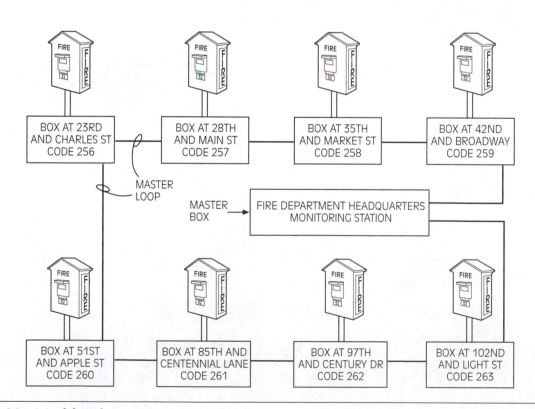

Figure M-14 Municipal fire alarm system.

National Pipe Hose Thread (NPSH) A standard fire hose thread that has dimensions for inside and outside screw threads.

National Standard Thread (NST) Threads used for hose and appliance screwed connections as specified in NFPA 1963. Also called a Standard Thread.

New York Pike Pole See *Hook*

Nozzle A device for directing a pressurized water stream in a desired pattern, density, and direction. Various nozzles are capable of projecting solid, heavy streams of water, curtains of spray, or fog (Figure N-1A). Nozzles have three main functions: they control flow, they provide shape, and they provide reach for firewater application. The flow of the water is controlled by the size of the nozzle's orifice. The nozzle itself creates a restriction at the end of the waterway that changes the water pressure into velocity, which allows the water to travel distances. There are four basic types of nozzles: 1) a smooth bore that has various solid or stacked tips; 2) the single gallonage (sometimes called variable pressure/variable flow); 3) the adjustable gallonage; and 4) the automatic or constant pressure. The following is a list of nozzles:

- **Cellar** A special nozzle (Figure N-1B) normally used to attack fires in the lower levels of a structure, such as basements and cellars (Figure N-1C). A Bresnan distributor can be attached to a cellar nozzle to attack cellar fires or fires in attics and other difficult-to-reach places (Figure N-1D).

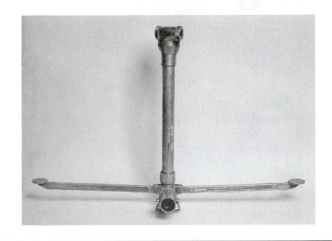

Figure N-1B Distributor pipe with a cellar nozzle attached.

Figure N-1C Cellar nozzle being used on a below-ground fire.

Figure N-1A Nozzles showing the various straight, solid, and wide-pattern streams.

Figure N-1D Bresnan distributor nozzles.

• **Combination** A manually adjustable nozzle capable of delivering water from a straight stream to a wide fog pattern (Figure N-1E). Also called a Variable Combination Fog Nozzle.

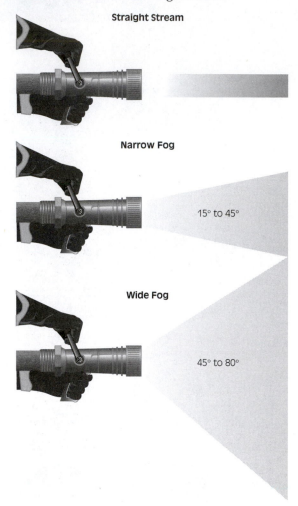

Straight Stream

Narrow Fog

15° to 45°

Wide Fog

45° to 80°

Figure N-1E Variable combination fog nozzle patterns.

• **Distributor** A firefighting spray nozzle normally used to combat basement-level fires. Also called a cellar nozzle or cellar pipe.

• **Distributor Pipe** A device that allows a nozzle or other device to be directed into a hole to reach basements, attic, and floors that cannot be accessed by personnel. The distributor pipe has self-supporting brackets that help hold it in place when in use. Also called an Extension Pipe.

• **Elevated** A water nozzle provided on an elevating or telescoping boom or platform of a fire apparatus to apply water at a higher level than normal grade.

• **Foam** Designed for low- and medium-expansion foams, usually having expansion ratios of 8:1 to 20:1 in the low range and up to 50:1 in the medium range (Figure N-1F). There are foam nozzles with

in-line and by-pass eductors (Figure N-1G), clip-on nozzle foam attachments (Figure N-1H), and medium-expansion foam generator nozzles (Figure N-1I). Also shown is a schematic of a foam generator nozzle (Figure N-1J). The bubble cup foam nozzle can be used like a fog nozzle with both Class A and B fires (Figure N-1K).

Figure N-1F Typical foam nozzles.

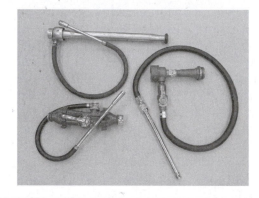

Figure N-1G Foam nozzles with in-line and by-pass eductors.

Figure N-1H Clip-on foam nozzle attachment.

Figure N-1I Medium-expansion foam generator nozzle.

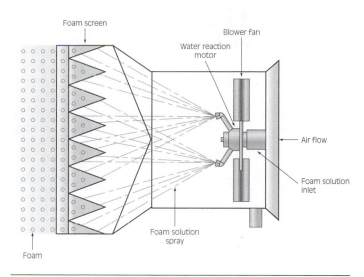

Figure N-1J Medium-expansion foam generator nozzle.

Figure N-1K Bubble cup foam nozzle.

- **Fog** Capable of producing a spray or water that contains water droplets with a mass medium diameter smaller than 0.03 inch in diameter (Figure N-1L). (A spray nozzle of water that contains water droplets with a mass medium diameter larger than 0.03 in. in diameter is technically considered a "spray" nozzle.) Shown is a schematic of a fog nozzle (Figure N-1M).

Figure N-1L Various styles of fog nozzles.

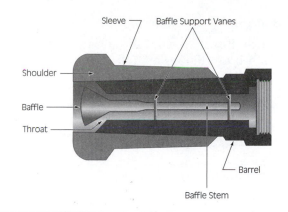

Figure N-1M Parts of a fog nozzle.

- **Forestry** This nozzle has two different types of tips built onto a single shut-off. If the handle is turned up the nozzle is a spray and if the handle is turned down then it becomes a straight stream nozzle (Figure N-1N). This nozzle is used with small hoses to conserve water while fighting wildland fires.

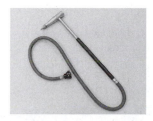

Figure N-1N Forestry nozzle.

- **Partition** Designed to discharge water between partition studs or joists in concealed building construction. Partition nozzles have a piercing point (Figure N-1O) or applicator head for penetrating through wall construction surfaces. Also called Penetrating or Piercing Nozzles.

Figure N-1O Partition nozzle.

- **Play Pipe** A nozzle without a shut off that is used for testing purposes only (Figure N-1P).

Figure N-1P Play pipe.

- **Revolving**   See *Cellar Nozzle*
- **Smooth Bore**   A straight stream nozzle with a smooth bore (Figure N-1Q).

Solid Stream

Figure N-1Q   Smooth bore nozzle.

- **Spray**   See *Fog Nozzle*
- **Straight Stream**   A nozzle that produces a straight stream of water that can travel long distances or penetrate a material to reach deep-seated fires. Shown is a schematic comparison of a straight stream and a solid stream at the tip (Figure N-1R).

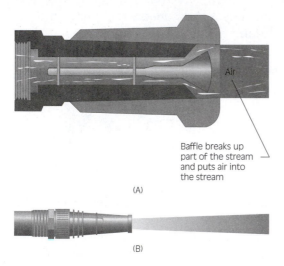

Baffle breaks up part of the stream and puts air into the stream

(A)

(B)

Figure N-1R   Comparison of (A) straight and (B) solid streams at tip.

- **Tip**   A solid stream or smooth bore nozzle that delivers an unbroken or solid stream of water at the tip. Tips can be stacked to adjust the size of the stream (Figure N-1S).

Figure N-1S   Various solid tips.

- **Water Curtain**   Designed to spray water to protect against exposure heat (Figure N-1T).

Figure N-1T   Water curtain nozzle.

Nozzle Shut-Off   There are two basic types of shut-offs—the lever type (Figure N-2A) and the rotating type (Figure N-2B) that can adjust the fog pattern or turn the nozzle on and off.

Figure N-2A   Various shut-off devices for nozzles.

Figure N-2B   Rotating nozzle shut-off.

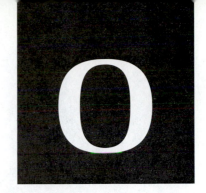

O-Tool See *Bar, Officer's Tool.*

Open-Circuit Breathing Apparatus See *Breathing Apparatus*

Ortho-Photo Map See *Map*

Outside Stem and Yoke (OS&Y) Valve A control valve for sprinkler and standpipe systems with a wheel on a stem housed in a yoke or housing (Figure O-1A). When the stem is exposed or outside, the valve is open (Figure O-1B). Also called an Outside Screw and Yoke Valve.

Orange Book See *DOT Emergency Guide Book*

Officer's Tool See *Bar*

Figure O-1A Outside stem and yoke (OS&Y).

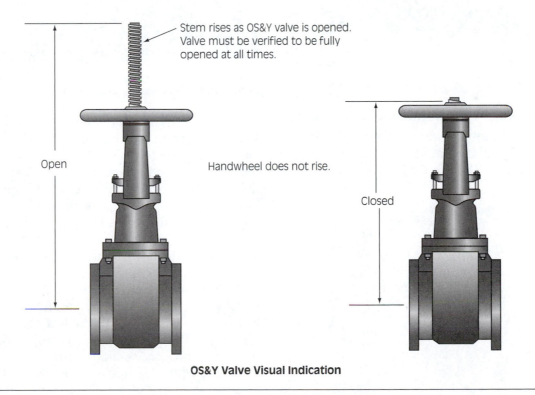

Stem rises as OS&Y valve is opened. Valve must be verified to be fully opened at all times.

Open

Handwheel does not rise.

Closed

OS&Y Valve Visual Indication

Figure O-1B Schematic of an OS&Y valve.

Patrol Unit Any light, mobile vehicle and crew with limited pumping and water capacity (Figure P-1).

Figure P-1 Patrol unit.

Pendant Sprinkler See *Sprinkler*

Personal Alert Safety System Device (PASS) A device that emits a loud alert or warning that the wearer is motionless (Figure P-2).

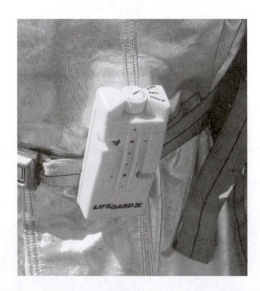

Figure P-2 PASS device.

Personal Flotation Device (PFD) A life vest or life preserver (Figure P-3).

Figure P-3 Personal flotation device (PFD).

Personal Protective Equipment (PPE) The NFPA has developed standards for personal firefighter protective equipment. Each type of operation requires a different type of protective clothing and equipment. For example, structure gear would be too heavy and bulky for wildland firefighting while wildland gear would be too light for structure firefighting. Protective equipment needs to be appropriate for the type of emergency to which the firefighter is responding. The specific NFPA standards listed in Figure P-4A primarily address the design, performance, and manufacturing of PPE. The NFPA has also developed an important PPE use guideline, called *Standards on Fire Department Occupational Safety and Health Program,* which dedicates a chapter to the use, care, and maintenance of many forms of protective clothing. All firefighting personnel must be equipped with proper equipment and clothing in order to mitigate the risk of injury from, or exposure to, hazardous conditions encountered while working.

• **Crash Fire Rescues** Firefighters that respond to aircraft fires wear proximity suits. Proximity gear

NFPA Standards That Address PPE and Ensembles

1500	Fire Department Occupational Safety and Health Program
1971	Protective Ensemble for Structural Firefighting
1975	Station/Work Uniforms for Firefighters
1976	Protective Clothing for Proximity Firefighting
1977	Protective Clothing and Equipment for Wildland Firefighting
1981	Open-Circuit Self-Contained Breathing Apparatus for the Fire Service
1982	Personal Alert Safety Systems
1983	Life Safety Rope and System Components
1991	Vapor-Protective Hazardous Ensembles/Materials Emergencies
1999	Protective Clothing for Medical Emergency Operations

Figure P-4A NFPA standards that address PPE.

utilizes an aluminized coating to help reflect radiant heat. Although similar in many ways to the structure PPE, the proximity gear must meet more stringent heat reflection and wearer insulation standards (Figure P-4B).

- **Hazardous Materials** There is no one set of hazardous materials PPE that can protective the firefighter from every situation. Given here are a few of the minimum protective equipment for hazardous materials responders.
  - **Level A Protective Clothing** This is the highest level of protection against chemical exposures. It is a fully encapsulated suit, therefore sometimes called an encapsulated suit instead of a Level A suit (Figure P-4C).
  - **Level B Protective Clothing** There are a variety of Level B suit. The two basic types are coverall style and encapsulated. The Level B suit is sometimes referred to as a bubble B or B-plus suit. The one factor that makes the Level B suit different from the lower levels of PPE is the addition of a self-contained breathing apparatus (SCBA). A Level A suit is gastight, whereas the Level B suit is intended for splash protection and, along with the SCBA, offers respiratory protection (Figure P-4D).

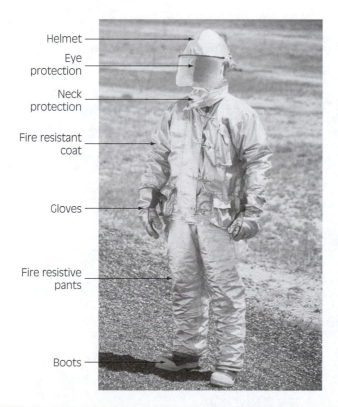

Helmet

Eye protection

Neck protection

Fire resistant coat

Gloves

Fire resistive pants

Boots

Hood

Figure P-4B Personal protective equipment (crash fire rescue).

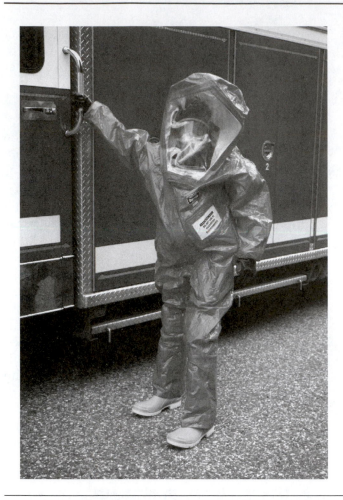

Figure P-4C   Level A protective clothing.

○ **Level C Protective Clothing**   Incorporates the use of an air-purifying respirator within the ensemble, used when splashes may occur, but when respiratory hazards are minimal and are covered by the use of an air-purifying respirator (Figure P-4E).

Figure P-4E   Level C protective clothing.

Figure P-4D   Level B suits—a coverall style, an encapsulated suit, and a two-piece suit.

○ **Level D Protective Clothing** Regular work clothing is used when respiratory protection is not required and splashes are not a concern. Level D suits provide no chemical protection, but do offer protection against other workplace hazards (Figure P-4F).

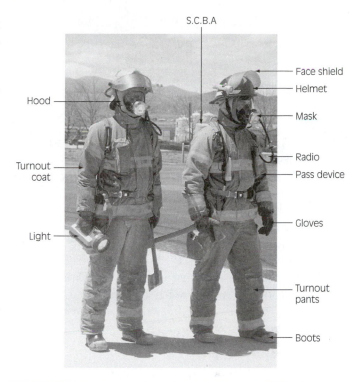

Figure P-4H   Personal protective equipment (structure).

Figure P-4F   Level D protective clothing.

• **Structure Fires** Structure firefighting includes rescue, fire suppression, and property conservation in buildings, enclosed structures, aircraft interiors, vehicles, vessels, or similar properties that are involved in a fire or emergency situation. A list of structure firefighting protective equipment is given in Figure P-4G, along with the typical structure firefighter and personal protective equipment (Figure P-4H).

Structural Firefighting PPE Ensemble Components

• Helmet
• Goggles
• SCBA
• Coat
• Pants
• Boots
• Hood
• Radio
• Flashlight
• PASS Device
• Pocket Tools
• Gloves

Figure P-4G   Structure firefighting PPE components.

• **Wildland Fires** Includes, but is not limited to: 8-inch, high-laced leather boots with lug soles, fire shelter, helmet with chin strap, goggles, ear plugs, fire-resistant shirt and trousers, leather gloves, drinking water, and individual first aid kits (Figure P-4I).

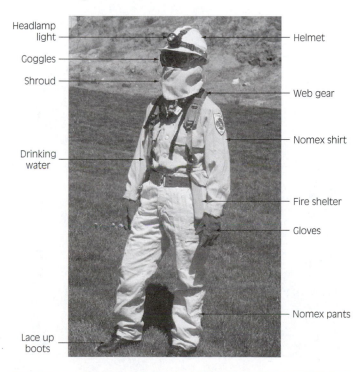

Figure P-4I   Personal protective equipment (wildland).

P

Pick-Head Axe   See *Axe*

Piercing Nozzle   See *Nozzle, Penetrating*

Pike Pole   See *Hook*

Ping Pong Ball Aerial Ignition Device   See *Aerial Ignition Device*

Pitot Tube   See *Gauge*

Placard   A first responder's hazardous materials identification system. The most common system for emergency responders is the Department of Transportation (DOT) placards (Figure P-5A). The quantity of hazardous materials that must be carried to require placarding is 1,001 pounds, unless it is a Table 1 material that requires placarding at any amount. Shown in Figures P-5B–O are the DOT placards found in the Department of Transportation guidebook that is carried with emergency response personnel.

- Class 1—Explosive (Figure P-5B)
- Class 2—Gases (Figure P-5C)
- Class 3—Flammable liquids (Figure P-5D)
- Class 4—Flammable solids (Figure P-5E)
- Class 5—Oxidizers (Figure P-5F)
- Class 6—Poisonous materials (Figure P-5G)
- Class 7—Radioactive materials (Figure P-5H)
- Class 8—Corrosives (Figure P-5I)

Figure P-5B   Class 1 explosive placards.

Figure P-5A   Hazardous materials placard on vehicle.

Figure P-5C   Class 2 gas placards.

Figure P-5D Class 3 flammable and combustible placards.

Figure P-5E Class 4 flammable solids placards.

Figure P-5F Class 5 oxidizers placards.

Figure P-5G Class 6 poisonous materials placards.

Figure P-5H Class 7 radioactive placards.

Figure P-5I Class 8 corrosives placards.

P

• Class 9—Miscellaneous hazardous materials (Figure P-5J)

Figure P-5J Class 9 miscellaneous hazardous materials placards.

• Harmful (Figure P-5K)

Figure P-5K Harmful placards.

• Hot (Figure P-5L)

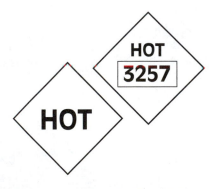

Figure P-5L Hot placards.

• Dangerous (Figure P-5M)

Figure P-5M Dangerous placards.

• Chlorine (Figure P-5N)

Figure P-5N Chlorine placards.

• Marine pollutant (Figure P-5O)

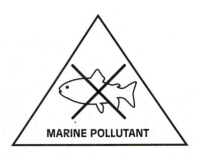

Figure P-5O Marine pollutant placards.

Plaster Hook See *Hook*

Play Pipe See *Nozzle*

Plug Tapered redwood or rubber in various sizes used for stopping leaks in tanks and pipes (Figure P-6).

Figure P-6 Plugs.

Pompier See *Ladder*

Pompier Belt See *Belt*

Portable Pump A small gasoline-powered pump that can be carried to a water source by one or two firefighters (Figure P-7).

Figure P-7   Portable pump.

Portable Radio   See *Radio*

Portable Tank   A container with either a rigid or self-supporting frame that can be filled with water or fire chemical mixture from which fire suppression resources can be filled (Figure P-8).

Figure P-9   Post indicator valve (PIV).

Figure P-8   Portable tank.

Positive Pressure Breathing Apparatus   See *Breathing Apparatus*

Post Indicator Valve (PIV)   Provides a visual means of indicating "open" or "shut" positions; found on the supply main of installed fire protection systems (Figure P-9).

Powered Hydraulic Cutter   A large rescue tool whose two blades open and close through the use of hydraulic power supplied through hoses from a power unit (Figure P-10) to cut through a variety of materials.

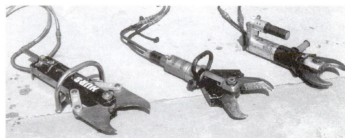

Figure P-10   Powered hydraulic cutters.

Powered Hydraulic Spreader   A large rescue tool whose two arms open and close through the use of hydraulic power supplied through hoses from a power unit, capable of exerting in excess of 20,000 pounds of force at their tips (Figure P-11).

Figure P-11 Powered hydraulic spreaders.

Powered Hydraulic Pump Provides the required fluid and pressure to operate a variety of spreaders, cutters, and rams (Figure P-12) and is powered by a gasoline engine, an electric motor, an air-driven motor, or the apparatus engine itself through a power take-off.

Figure P-12 Powered hydraulic pumps.

Powered Hydraulic Ram Designed to push and pull heavy objects; however, the pulling strength is generally one half that of the pushing strength (Figure P-13).

Figure P-13 Powered hydraulics rams.

Proportioner A device to introduce the correct amount of agent (especially foam and wetting agents) into streams of water. See also *Foam Applicator* and *Balanced-Pressure Demand-Type Proportioning System*

Psychrometer The general name for instruments designed to determine the moisture content of air. A psychrometer consists of dry and wet bulb thermometers that give the respective temperatures, which in turn are used to determine relative humidity and dew point (Figure P-14).

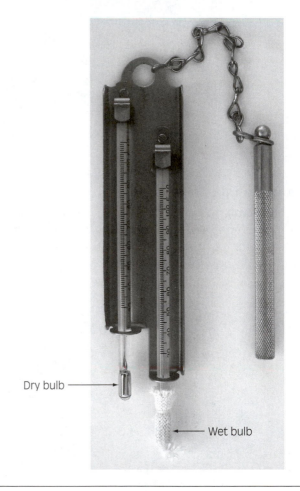

Dry bulb

Wet bulb

Figure P-14 Psychrometer.

Pulaski A combination chopping and trenching tool that combines a single-bit axe blade with a narrow adze-like trenching blade fitted to a straight handle. Useful for grubbing or trenching in duff and matted roots (Figure P-15A). Shown is a schematic of the Pulaski (Figure P-15B) and the sharpening procedures.

Pulley A device that transfers applied force from one location to another when used in conjunction with a flexible medium, such as a rope, chain, belt,

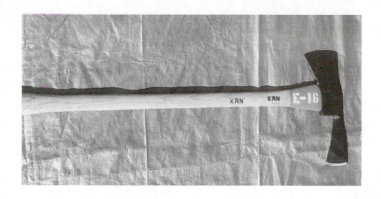

Figure P-15A Pulaski axe.

Figure P-16A Double pulley with becket.

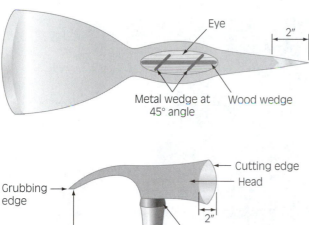

Eye
2″
Metal wedge at 45° angle   Wood wedge

Cutting edge
Head
Grubbing edge
2″
3/8″ bevel at 45° angle
1/2″ shoulder

Handle

Figure P-15B Pulaski parts and sharpening procedures.

Figure P-16B Pulleys of different sizes and shapes.

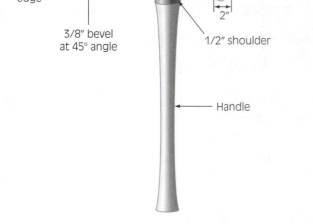

Center plate
a
a
b
b
c
c
d

a – Side plates
b – Sheave(s)
c – Axle
d – Becket (on double pulleys)

and so on (Figure P-16A). There are several pulleys of different shapes and sizes (Figure P-16B) used by the fire service. Shown is a drawing of a pulley and its parts (Figure P-16C).

Pump A fire pump is a device used to raise, transfer, or compress liquids or gases. Pumps used for gases and vapors are known as compressors. The minimum rated capacity of a mounted fire pump is not less than 750 gallons per minute (GPM). Pumps of higher capacity are rated in 250-GPM increments,

Figure P-16C A schematic view of a pulley.

considered to be a standard hose line stream. Pumps that deliver 1,000, 1,250, or 1,500 GPM are popular in the municipal fire service. The pump must be of centrifugal design with a stainless steel shaft and bronze impellors (Figure P-17A). Following are the basic types of pumps used in the fire service today:

- **Centrifugal Pump** Uses the tendency of a body of water to move away from the center when rotating in a circular motion. Centrifugal pumps will not pump air and therefore need a primer (Figure P-17B). A centrifugal pump uses an impeller to propel water at a rate according to the speed of the pump. Shown here is the housing of the centrifugal pump impeller (Figure P-17C), along with a cutaway of a single-stage centrifugal pump (Figure P-17D).

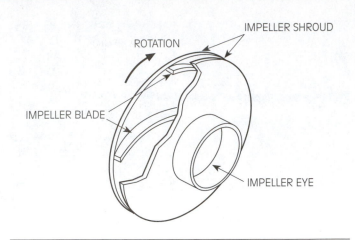

Figure P-17C Centrifugal pump impeller.

Figure P-17D Cutaway of a single stage centrifugal pump showing the impeller on a shaft.

Figure P-17A Fire pump cutaway.

- **Floating Pump** Small, portable pump that floats on water (Figure P-17E).

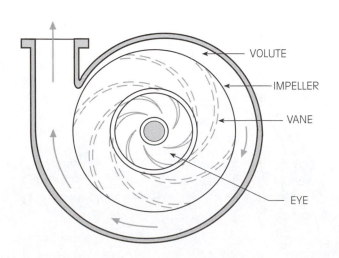

Figure P-17B Interior design of centrifugal pump.

Figure P-17E Floating pump.

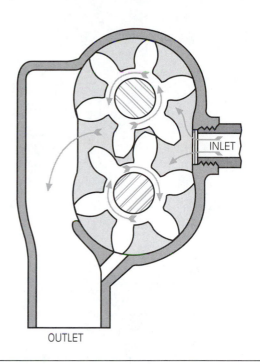

Figure P-17F Gear pump: a) interior components; b) exterior view.

• **Positive Displacement Pump** Moves a specified quantity of water through the pump chamber with each stroke or cycle; it is capable of pumping air, and therefore is self-priming, but must have a pressure relief provision if plumbing or hoses have shut-off nozzles or valves. Examples of positive displacement pumps are the gear pump (Figure P-17F), and the piston pump (Figure P-17G), the rotary lobe pump (Figure P-17H), and the rotary vane pumps (Figure P-17I).

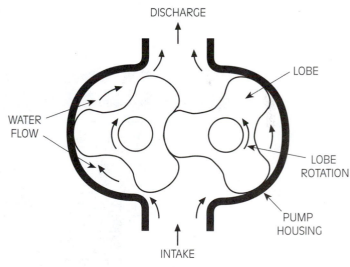

Figure P-17H A rotary lobe pump showing the flow of water from the intake to the discharge side of the pump.

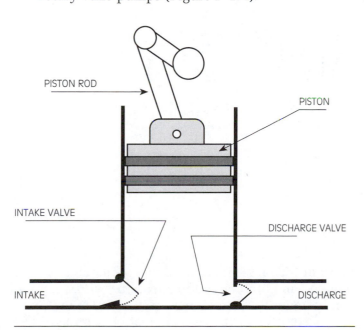

Figure P-17G Piston or reciprocating pump.

• **Priming Pump** Any of several small positive displacement pumps used to prime a centrifugal pump by removing the air to let water flow into the pump (Figure P-17J). Shown is a schematic of a priming valve (Figure P-17K).

• **Pump Can** See *Backpack Pump*

• **Pump Panel** The instrument and control panel located on the pumper (engine) (Figure P-17L).

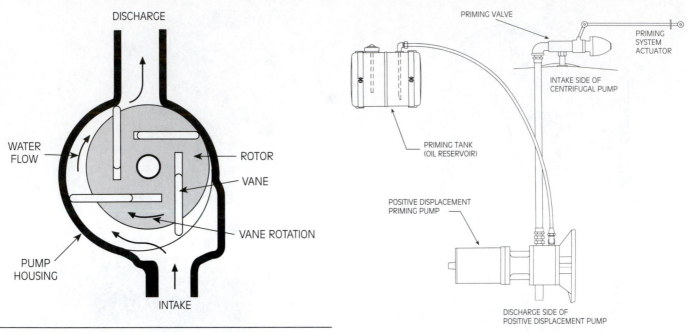

Figure P-17I A rotary vane pump showing the flow of water from the intake to the discharge of the pump.

Figure P-17J Components of a typical priming system.

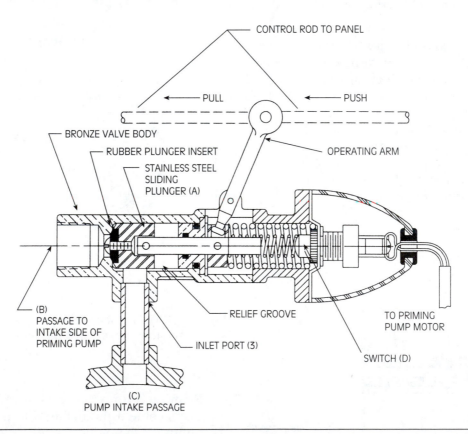

Figure P-17K Priming valve.

- **Pump Transfer** A centrifugal water pump that is mounted on the apparatus and is used for water transfer from its tank. The pump can be driven either by a separate engine or by a power take-off from the apparatus transmission (Figure P-17M).

Pumper The basic unit of fire apparatus is the pumper, an automotive fire apparatus with a permanently mounted fire pump, water tank, and hose body also called a "triple-combination pumper." The unit is designed for sustained pumping operations

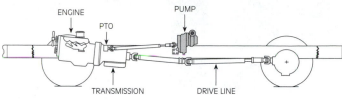

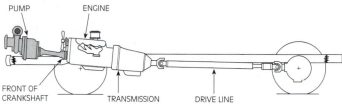

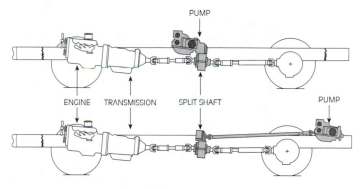

Figure P-17L   Pump panel.

Figure P-17M   Pump transfer.

during structural firefighting and is capable of supporting associated fire department operations. The vehicle also may be equipped with an optional water tower to provide an elevated master stream for fire suppression (Figure P-18). See also *Engine*

- **Class A** Will deliver its rated capacity of 150 psi net pump pressure at a lift of no more than 10 feet with a motor speed of not more than 80 percent of the certified peak of the brake horsepower curve; will deliver 70 percent of rated capacity at 200 psi and 50 percent of rated capacity at 250 psi. Also called an Engine.

- **Class B** Will deliver its rated capacity of 120 psi net pump pressure at a lift of no more than 10 feet and a motor speed not exceeding 80 percent of the certified peak of the brake horsepower curve;

will deliver 50 percent of its rated capacity at 200 psi and 33.33 percent of its rated capacity at 250 psi. Class B pumps have not been manufactured since the mid-1950s.

P

Figure P-18 Pumper or engine has the ability to carry firefighters, tools, equipment, and hose necessary for fire suppression operations.

Quad A quadruple fire apparatus equipped with a fire pump, water tank, hose bed storage, and the addition of a full complement of ground ladders. This type of apparatus is most useful in areas where a ladder company does not exist or can be expected to be delayed (Figure Q-1).

Figure Q-1 Quad.

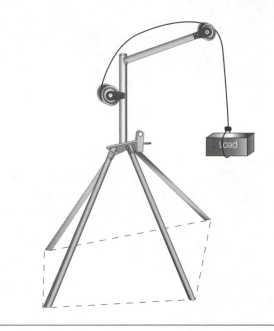

Figure Q-2 Quad pod.

Quad Map A topographical map referred to as quadrangles or quads.

Quad Pod A retrieval support device consisting of a davit arm centered over four legs that support the device (Figure Q-2).

Quint A quintuple fire apparatus equipped with a fire pump, water tank, hose bed storage, a full complement of ground ladders, and an aerial device (Figure Q-3).

Figure Q-3 Quint.

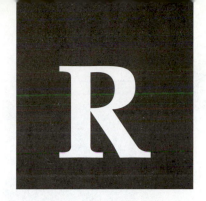

Radio Chest Harness   A harness worn on the chest that holds a portable radio (Figure R-1).

Figure R-1   Radio chest harness.

Figure R-2A   Fixed radio or radio base unit.

Figure R-2B   Mobile radio.

Radio   The primary link between the communications center and field units. There are basically four radio systems:

- **Fixed Radio**   Includes transmitters, control consoles, fixed receivers, and satellite equipment. Also called Base Units (Figure R-2A).
- **Mobile Radio**   A radio and equipment that is installed in a vehicle (Figure R-2B).

- **Paging Unit**   An electronic device used to contact people via a paging network (Figure R-2C). It predates mobile phone technology, but similarly uses radio transmissions to communicate between a control/call center and the recipient.
- **Portable Radio**   Handheld battery-powered radio (Figure R-2D).

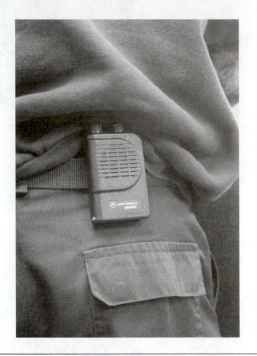

Figure R-2C Pager.

Rake Used by firefighters to clear away vegetation. Shown here is the asphalt rake (Figure R-3). See also *Council Rake, Fire Rake,* and *California Barron Tool.*

Ramp See *Hose Tool*

Reciprocating Saw See *Saw*

Reducer Coupling See *Hose Coupling*

Reel See *Hose Tool*

Relay Tank See *Portable Tank*

Relief Valve See *Hose Appliance*

Figure R-2D Portable radio.

Figure R-3 Asphalt rake, also called a fire rake.

Repeater Receives radio transmissions and boosts the signal. Used in areas where topography interrupts clear communications (Figure R-4).

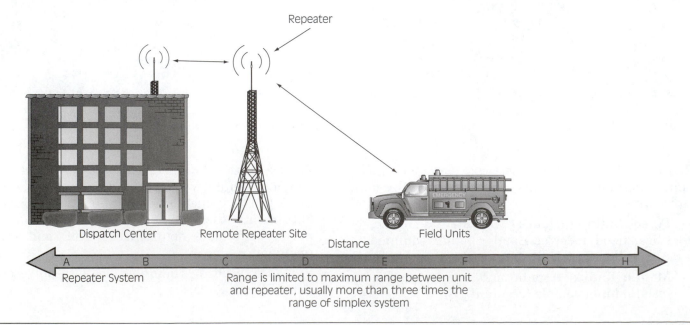

Figure R-4 A repeater extends radios' operating range.

Rescue Company A rescue truck and its firefighters (Figure R-5). A rescue company is equipped and trained to handle a variety of duties including search and rescue, medical treatment, fire suppression, and the extrication of victims in motor vehicle accidents. The actual duties of a rescue company can vary in different parts of the country, as does the term to describe one. For example, a rescue company is called a squad in some areas, while other areas use that term to refer to their ambulances.

Figure R-5 Rescue company.

Rescue Knife A V-shaped knife made of high-strength aluminum alloy used to cut seatbelts and webbing up to 10,000-pound strength (Figure R-6).

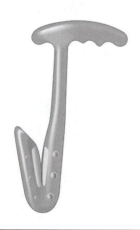

Figure R-6 Rescue knife.

Retard Chamber A small tank attached to a sprinkler system that allows pressure surges to dissipate their energy before they enter the system and set off the water flow alarm (Figure R-7).

Rex Tool A lock-pulling tool shaped like a "U" with sharp, tapered blades that bite into lock cylinders (Figure R-8).

Rigging Plate A device to keep ropes and equipment from becoming entangled (Figure R-9).

Figure R-7 Retard chamber.

Figure R-8 Rex tool.

Figure R-9 Rigging plates.

Riser A vertical water pipe used to carry water for fire protection systems above grade, such as a standpipe, or sprinkler riser (Figure R-10).

Roof Ladder See *Ladder*

Roofman's Hook See *Hook*

R

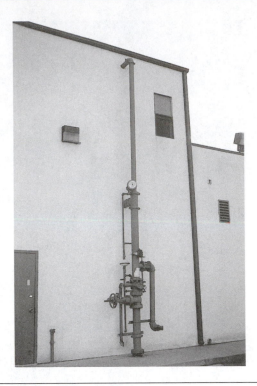

Figure R-10   Riser.

Rope   In the fire and rescue service, rope is divided into two categories—utility and life safety. A utility rope is only used for utility purposes and has no standard governing materials, required strength, safety factors, number of uses, or associated hardware (Figure R-11A).

A life safety rope is used for such purposes as rescue (Figure R-11B). One- or two-person rope requires

Figure R-11B   Life safety rope.

a minimum tensile strength of 4,500 pounds, while a two-person rope requires a minimum tensile strength of 9,000 pounds.

Figure R-11A   Utility rope: (a) hoisting a pike pole; (b) hoisting a ladder; and (c) hoisting a charged hose line.

Figure R-12 Parts of a rope.

Rope has three basic parts. The end of the rope used to tie the knot is called the *working end*. The length between the working end and the running end is called the *standing part*. The *running end* is used for work such as hoisting tools (Figure R-12).

NFPA 1983 deals with life safety ropes, harnesses, and hardware. NFPA 1983 established minimum standards for this type of equipment if it is to be used by firefighters during the performance of their duties.

Rope is constructed of a variety of materials. Each material has a different set of characteristics for utilization in the fire rescue service. Natural fiber ropes are constructed using the laid (twisted) (Figure R-13A) method and synthetic ropes are either braided (woven) (Figure R-13B), laid braid on braid

(braiding a sheath over a smaller braided core) (Figure R-13C), or made in the kernmantle style, (Figure R-13D) which has an outer layer called *kern* and an inner layer called *mantle.*

Braided

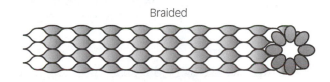

Figure R-13B Rope construction—braided.

Braid-on-Braid

Figure R-13C Rope construction—braid-on-braid.

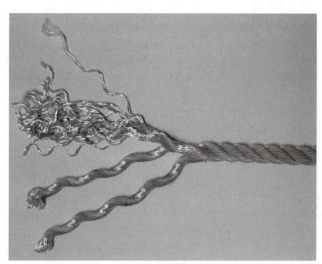

Figure R-13A Rope construction—laid or twisted.

Figure R-13D Rope construction—kernmantle.

1. **Natural Materials** These ropes have low resistance to rot, mildew, abrasion, natural deterioration/degradation due to age, along with a very low strength-to-weight ratio and a low shock load absorption capacity.

   • **Manila** Made from fibers that grow in the leafstalk of the abaca plant. Manila ropes are rarely used in the fire service due to their loss in strength when they get wet (Figure R-14A).

Figure R-15A Nylon rope.

Figure R-14A Manila rope.

   • **Sisal** A fiber obtained from plant leaves. Sisal ropes have approximately 25% less tensile strength than manila fiber ropes of a similar diameter.

   • **Cotton** Made from the seed hairs of the cotton boll. This rope is soft, pliable, and easily damaged, with 50% less tensile strength than manila rope.

2. **Synthetic Materials** These ropes have high resistance to rot, mildew, and natural degradation due to age. They are also more resistant to physical damage and damage from abrasion than natural material ropes. Synthetic ropes are stronger and more durable because their fibers are continuous from end to end, unlike natural fiber ropes that are made up of short fiber strands.

   • **Nylon** Made of the same material as fishing line. Nylon ropes are constructed the same way as multifilament bundles. Nylon is susceptible to damage by acids and has a loss of up to 25% strength when wet, frozen, and stretched under loads (Figure R-15A).

   • **Polypropylene** Ropes constructed from this material are primarily used for water rescue, as water has no effect on their strength and because they float. This rope has a high resistance to rot and mildew but is susceptible to damage from sunlight and has a relatively low breaking point (Figure R-15B).

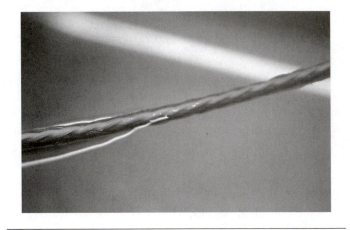

Figure R-15B Polypropylene rope.

   • **Polyethylene** This rope floats and comes in highly visible colors, with similar properties as polypropylene (Figure R-15C).

Figure R-15C Polyethylene rope.

R

Mayberry Fire and Rescue Department Life Safety Rope Inspection and History Log						
Date	Location	Type of use	Sheath fray, %	Other damage	Additional comments	Inspected by

Figure R-16 Rope log.

- **Polyester** This rope has high resistance to both acids and alkalis, with low elongation under load. Its strength is unaffected by being wet or frozen, although polyester rope does not handle shock loading well.

Rope Hose Tool See *Hose Tool*

Rope Log A form to keep track of a rope's history (Figure R-16).

Rope Washer 1) A small PVC device that fastens directly to a hose bib or to the male end of a garden hose (Figure R-17A). 2) Kernmantle rope can be placed in a mesh bag and washed in a front-loading washing machine (Figure R-17B).

Figure R-17B Front-loading rope washer.

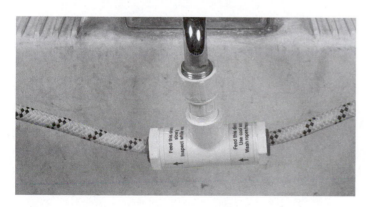

Figure R-17A Rope washer (PVC).

Rotary Power Saw See *Saw*

Rotary Vane Pump See *Pump*

Rubber-Covered Hose See *Synthetic Nitrile Rubber Hose*

Run Card System Provides specific information on what apparatus and personnel respond to specific areas of a jurisdiction using cards or another form of documentation (Figure R-18).

Rung See *Ladder Construction*

Runner See *Floor Runner*

R

BLOCK	ENGINES	TRUCKS	RESCUE	D/CHIEF	A/CHIEF	MEDIC	OTHER
1000	1, 5, 4, 7, 10	1, 7, 10	1, 9	1, 2, 3	1	1, 4, 5	
1100	1, 5, 4, 7, 10	1, 7, 10	1, 9	1, 2, 3	1	1, 4, 5	
1200	1, 5, 4, 7, 10	1, 7, 10	1, 9	1, 2, 3	1	1, 4, 5	
1300	1, 4, 5, 7, 10	1, 7, 10	1, 9	1, 2, 3	1	1, 4, 5	
1400	1, 5, 4, 7, 10	1, 7, 10	1, 9	2, 1, 3	1	1, 5, 4	
1500	1, 5, 4, 7, 10	1, 7, 10	1, 9	2, 1, 3	1	1, 5, 4	

STREET NAME: Texas Ave. BOX NUMBER: 1128
BLOCK RANGE FROM: 1000 TO: 1500

STREET NAME: Texas Ave BLOCK RANGE FROM: 1000 TO: 1500

Figure R-18 Run card.

Salvage Basket Usually made of canvas or synthetic material for salvage and overhaul. Also called a debris bag (Figure S-1). See also *Salvage Bucket* and *Salvage Bag*

Figure S-1 Salvage basket.

Salvage Cover A tarp made of canvas or plastic (Figure S-2A) used to cover furniture during salvage operations (Figure S-2B). It can also be used to make water chutes and catch basins for water run off. See also *Visqueen*

Figure S-2A Salvage cover.

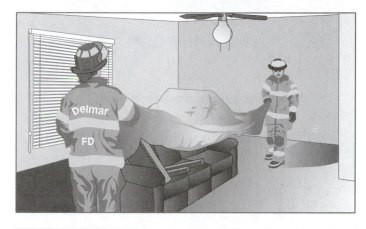

Figure S-2B Salvage cover deployment.

Salvage Vacuum A vacuum that picks up water, with a limited capacity of 5 gallons. One style of salvage vacuum is worn like a backpack (Figure S-3) and the other is moved around on wheels.

Figure S-3 Salvage vacuum.

San Francisco Bar   See *Bar*

Saw   Firefighters use two types of saws—handsaws and power saws.

1. Hand Saws

- **Hack Saw**   A metal cutting saw with strong blades; mainly used for vehicle rescues (Figure S-4A).

Figure S-4A   Hack saw.

2. Power Saws

- **Chain Saw**   A portable power saw with a continuous chain that carries the cutting teeth. Chain saws are used for wildland (Figure S-5A) and structure firefighting (Figure S-5B) and may also be used in rescue operations.

Figure S-5A   Wildland chain saws.

Figure S-5B   Structure firefighting chain saw.

- **Rotary Saw**   A circular saw that has several types of blades for a variety of uses, such as cutting through security gates and overhead doors and for ventilation and vehicle extrication. Shown here is a wood saw (Figure S-5C) and masonry and metal saw (Figure S-5D).

Figure S-5C   Rotary saw with wood cutting blade.

Figure S-5D   Rotary saw with an abrasive disc for masonry or metal.

- **Reciprocating Saw**   A saw with a straight blade mounted at the end of a gun-like body (Figure S-5E). The blade moves back and forth (hence the name "reciprocating"), much like the action of a jigsaw, but reciprocating saws are much more powerful and versatile than jigsaws. If fitted with the correct blade they are able to cut through virtually anything but make rather rough cuts compared with other types of saws. Firefighters use reciprocating saws for extrication and salvage operations.

Self-Contained Breathing Apparatus See *Breathing Apparatus*

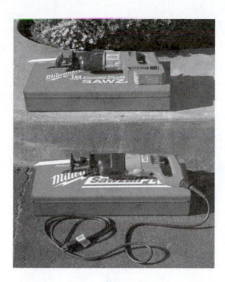

Figure S-5E Reciprocating saw.

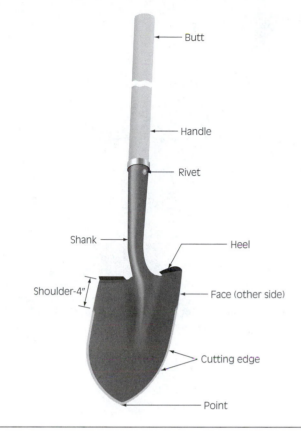

Figure S-7A Round point shovel for wildland firefighting.

Shoring The use of timbers to support and/or strengthen weakened structural members (roofs, floors, walls, and so on) in order to avoid a secondary collapse during rescue operations (Figure S-6).

Shove Knife A forcible entry tool used for interior doors with a key in the knob locks. See *Officers Tool to see Shove Knife under Bars Chapter B*

Shovel A tool for lifting and moving loose material such as snow, dirt, or sand. It has a broad blade with edges or sides that are fixed to a medium-length handle. Wildland firefighters use a long-handled round point shovel (Figure S-7A), probably the most universal tool for that type of fire service. It can be used to cut brush and grass, cut a fire line and throw dirt. The edges of the shovel should be sharpened for best use (Figure S-7B). Structure firefighters also use the square point shovel or scoop shovel (Figure S-7C).

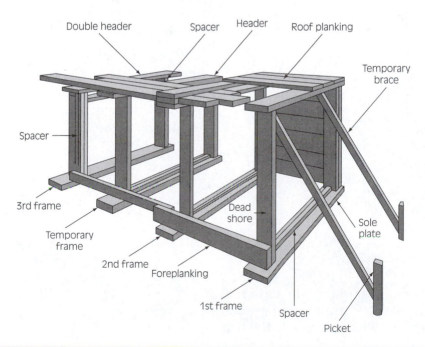

Figure S-6 Shoring.

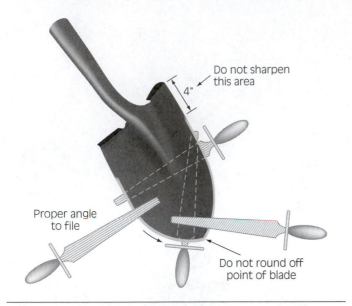

Figure S-7B Sharpening procedures for a wildland firefighting round point shovel.

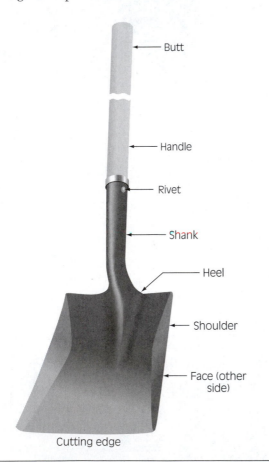

Figure S-7C Square point shovel.

Slim Jim A tool that opens car doors when inserted between the glass and weather stripping and pushed down on the control arm or when pushed to the rear and lifted (Figure S-8). Also called a Slip Lock Device or Door Opening Tool.

Figure S-8 Slim jim.

Smoke Detector See *Detector*

Smoke Ejector See *Blower*

Smoke Jumper A specially trained and certified firefighter who travels to wildland fires by aircraft and then parachutes to the fire.

Smooth Bore Nozzle See *Nozzle*

Snap Link See *Anti-Torque Device*

Snatch Block A pulley-block that can be opened to receive the bight of a rope, used for mechanical advantage (Figure S-9).

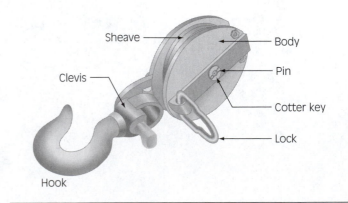

Figure S-9 Snatch block.

Snorkel A vehicle specially equipped with a hydraulically operated elevating boom mounted on a turntable, for use in either firefighting or rescue operations (Figure S-10). Also called a Snorkel Truck.

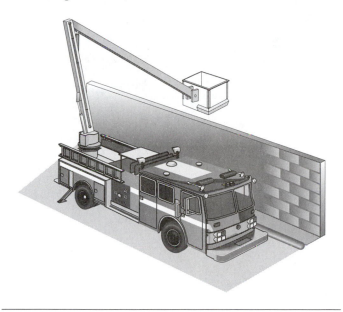

Figure S-10 A snorkel basket can reach places that are inaccessible with other types of apparatus.

Spanner Wrench A tool used to tighten or loosen couplings or turn hydrants on and off (Figure S-11A). It may also be used as a pry bar, door chock, and for turning off gas valves.

- **Booster Spanner** Also called a Pin Hole Spanner (Figure S-11B).
- **Hydrant Spanner** Includes the adjustable (Figure S-11C), five-hole pocket (Figure S-11D), and pocket combination (Figure S-11E) spanners.

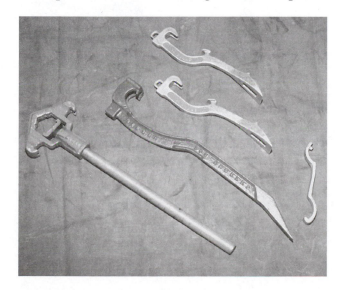

Figure S-11A Spanner wrenches.

Figure S-11B Booster hose or pin hole spanner.

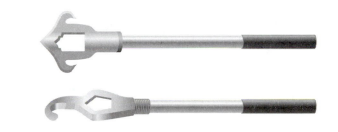

Figure S-11C Two types of adjustable hydrant wrenches.

Figure S-11D Five-hole hydrant wrench.

Figure S-11E Pocket combination hydrant spanner.

- **Rocker Lug Spanner** Includes the pocket or slotted (Figure S-11F) and forestry lug (Figure S-11G) spanners.
- **Storz Spanner** For large diameter hose (Figure S-11H) and for locking storz coupling (Figure S-11I).

S

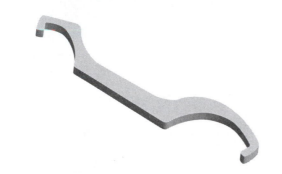

Figure S-11F   Pocket lug or slotted spanner.

Figure S-11G   Forestry lug spanner.

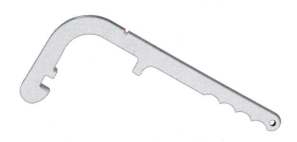

Figure S-11H   Storz spanner.

Figure S-11I   Storz spanner for locking storz coupling.

- **Universal Spanner** Can do several different tasks, such as prying windows, shutting off gas, and functioning as a hammer or spanner (Figure S-11J).

Sprinkler A water deflector spray nozzle used to provide distribution of water in specific characteristic patterns and densities for the purpose of cooling exposures to unacceptable heat radiation and controlling and suppressing fires or combustible vapor dispersions. Shown here are the parts of a sprinkler head

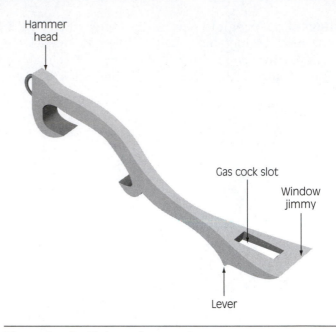

Figure S-11J   Universal spanner.

(Figure S-12A), along with different sprinkler designs (Figure S-12B). There are several types of sprinklers: The upright pendant (without a guard) and pendant sidewall (with guard), the old-style sprinkler, spray sprinkler, the standard response sprinkler, and the quick response sprinkler (Figure S-12C).

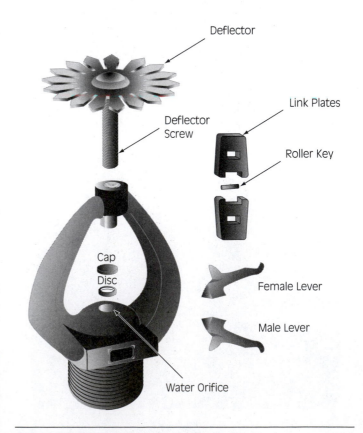

Figure S-12A   Sprinkler head parts.

Figure S-12B Assorted sprinklers.

Sprinkler Connection A fire department Siamese connection that allows pumpers (engines) to supplement the supply (Figure S-13).

Sprinkler Riser The vertical portion of a sprinkler system piping from the ground main to the horizontal cross main that feeds the branch lines (Figure S-14).

Sprinkler Stop A small device, either a tong or wedge, used to stop the flow of water from activated sprinkler heads (Figure S-15A). It is wedged between the orifice and bail of sprinkler heads or inserted into the orifice of sprinkler heads (Figure S-15B).

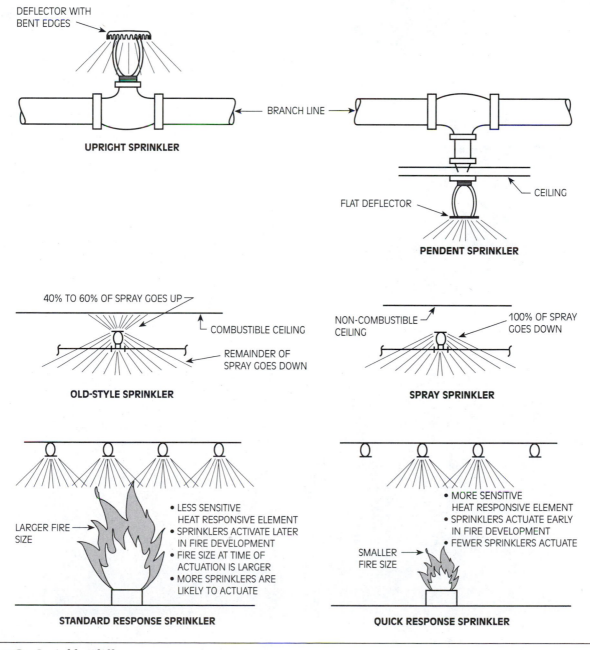

Figure S-12C Sprinkler differences.

Figure S-13   Fire department sprinkler connection.

Figure S-14   Sprinkler riser.

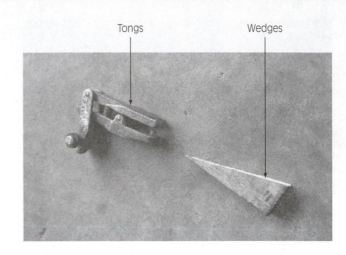

Figure S-15A   Sprinkler stops—tongs and wedges.

Squeegee   A tool with a flat, smooth, rubber blade used to remove or control the flow of liquid on a flat surface. Firefighters use squeegees to remove water after a fire or when a building is flooded after a sprinkler head goes off (Figure S-16).

Standard Coupling   Fire hose coupling with American National Standard (NH) threads.

Standpipe System   Piping system that allows for the manual application of water in large buildings. Each floor of the building will have a standpipe outlet for fire hose connections (Figure S-17A).

Standpipe Hose Cabinet   A cabinet for standpipe outlets and/or fire hose storage (Figure S-17B).

Figure S-15B   Tongs and wood wedges stopping sprinkler flow.

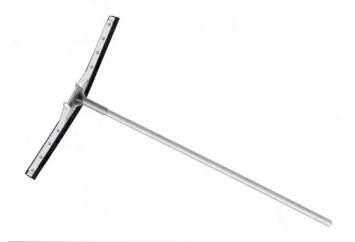

Figure S-16 Squeegee.

Figure S-17A Standpipe outlet for fire hose.

Figure S-17B Standpipe hose cabinet.

Stay Poles The stabilizer poles attached to the sides of Bangor ladders that are used to raise the ladder. Once raised, they are not used to support the extended ladder (Figure S-18).

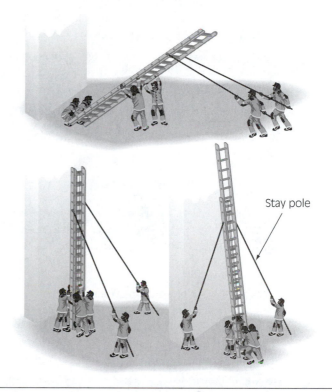

Figure S-18 Stay poles are only used for raising the ladder.

Storz Coupling See *Hose Coupling*

Straight Ladder See *Ladder*

Strainer See *Hose Appliance*

Stream Straightener A metal tube, commonly with metal vanes inside it, between a master stream appliance and its solid tip. The purpose is to reduce turbulence in the stream, allowing it to flow straighter.

Street Broom A wire broom intended for use on railways and street car switches, used by firefighters to clean debris and materials from pavement and hard surfaces (Figure S-19).

Figure S-19 Street broom.

Strike Team Specified combinations of the same kind and type of resources with common communications and a leader. For example, an engine strike team consists of five engines and a leader; a dozer strike team includes two dozers, a leader, and a service unit; and a crew strike team is composed of thirty personnel including a leader (Figure S-20).

Crew strike team: 30 members with leader

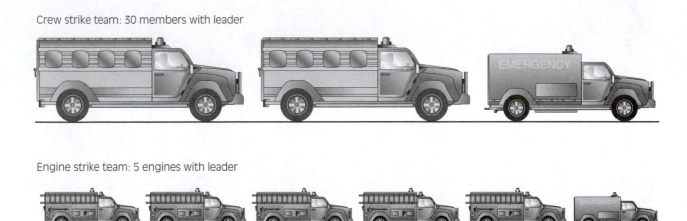

Engine strike team: 5 engines with leader

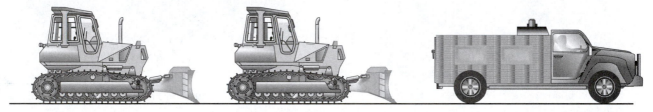

Dozer strike team: 2 dozers and a service unit with leader

Figure S-20 Strike teams—engines, crews, and dozers.

S

Tag Line   Ropes held and controlled by firefighters on the ground or lower elevations in order to keep items being hoisted from banging against or getting caught on the structure as they are rising (Figure T-1).

Figure T-1   Tag line.

Tanker   Aircraft capable of carrying and dropping fire retardant or water. Also called Water Tender Tankers. See also *Aircraft*

Task Force   Any combination of single resources assembled for a particular tactical need, with common communications and a leader. A task force may be pre-established and sent to an incident or formed at the incident (Figure T-2).

Tender   See *Water Tender*

Thermal Imaging Camera   A device that helps firefighters see in low-visibility environments (figure of a thermal imagining camera helmet mounted Figure T-3 and hand held thermal imaging camera Figure T-4).

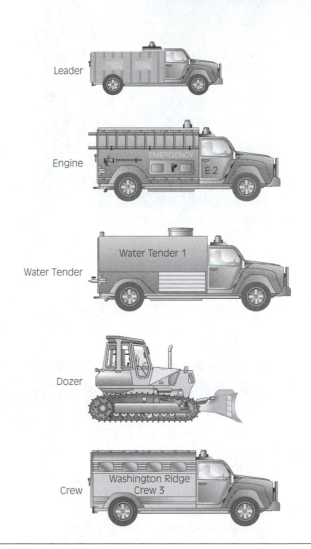

Figure T-2   A task force can be composed of any combination of single resources with a leader and common communications.

Figure T-3   Thermal imaging camera.

(Photo courtesy of SAFE-IR)

Figure T-4   Thermal imaging camera (TIC).

Tiller   A ladder truck in which the rear wheels are steered from the back of the truck (Figure T-5).

Figure T-5   Tiller.

Tiller Bucket   The area from which the tiller driver operates the rear steering of a tiller truck (Figure T-6).

Figure T-6   Tiller bucket.

Tip   See *Nozzle*

TNT Tool   A combination of several tools that can be used as a sledgehammer, axe, pike pole, or pry bar (Figure T-7). Also known as a Denver Tool.

Figure T-7   TNT tool.

Tower Ladder   A telescoping aerial platform of a fire apparatus (Figure T-8).

Figure T-8   Tower ladder.

Tractor Plow   Any tractor with a plow (Figure T-9) for constructing a fire line by exposing mineral soil. When assigning a type or level, a tractor plow includes the transportation and personnel for its operation.

Figure T-9   Tractor plow.

Transfer Pump   See *Pump Transfer*

Triple Combination Engine   A fire apparatus that can carry water, hose, and equipment and that can pump water (Figure T-10).

Figure T-10   Triple combination engine.

Tripod   Three-legged retrieval support device (Figure T-11).

Figure T-11   Tripod.

Trolley   A wheeled device that rolls along the track line and carries the weight of the load vertically and the force of the control lines horizontally (Figure T-12).

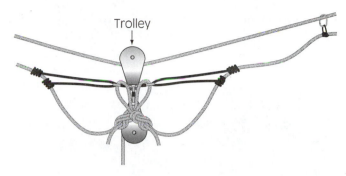

Figure T-12   Trolley.

Truck Company   A group of firefighters assigned to perform tactics and functions such as forcible entry, search and rescue, ventilation, and so on (Figure T-13).

Figure T-13   Truck company firefighter.

Turnout Clothing   Protective firefighters' clothing that meets the guidelines of NFPA standard 1971 (Figure T-14).

Figure T-14   Turnout clothing.

T

Turntable The rotating platform of a ladder that affords an elevation ladder device the ability to turn to any target from a fixed position (Figure T-15).

Figure T-15 Turntable.

Turret Pipe A large master stream appliance normally mounted on an engine that is connected to a pump's discharge (Figure T-16).

Figure T-16 Turret pipe.

Type Refers to resource capabilities. A Type 1 resource provides a greater overall capability due to power, size, and so on than would be found in a Type 2 resource. Resource typing provides managers with additional information in selecting the best resource for the task.

T

U

UHL Tool A heavy-duty forcible entry and overhaul tool that has a pick axe, pry bar, nail puller, gas shut-off, and ceiling hook (Figure U-1).

Figure U-1 UHL tool.

Universal Spanner Wrench See *Spanner Wrench*

Unlined Fire Hose See *Hose*

Upright Sprinkler See *Sprinkler*

Urban Search and Rescue (US&R) Crew A predetermined number of individuals who are supervised, organized, and trained principally for a specified level of US&R operational capability. They respond without equipment and are used to relieve or increase the number of US&R personnel at the incident.

Urban Search and Rescue (US&R) Company Any ground vehicle providing a specified level of US&R operational capability, rescue equipment, and personnel.

Utility Rope See *Rope*

Very Pistol *See Flare Launcher*

Vest High-visibility clothing that identifies users such as the Incident Commander, Safety Officer, and so on. Also worn to be highly visible in high-traffic areas (Figure V-1).

Visqueen A plastic sheet that can be used for salvage and overhaul operations (Figure V-2).

Figure V-1 Safety officer wearing a vest.

Figure V-2 Visqueen.

Wall Fire Hydrant   See *Hydrant*

Wall Pick   A pick hammerhead tool used to break up concrete, asphalt, or masonry during a rescue, forcible entry, overhaul, or salvage operation (Figure W-1).

Figure W-1   Wall pick.

Water Cannon   A large water application nozzle that cannot be handheld due to its size and water reaction (Figure W-2).

Figure W-2   Water cannon.

Water Curtain Nozzle   See *Nozzle*

Water Flow Alarm   A device that indicates the flow of a fluid and alerts occupants that the system has been activated or has been damaged and water flow is occurring, requiring immediate action. Shown here are the pressure switch (Figure W-3A) and vane (Figure W-3B) alarms.

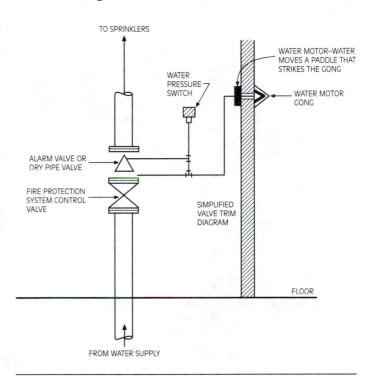

Figure W-3A   Pressure switch type of water flow alarm.

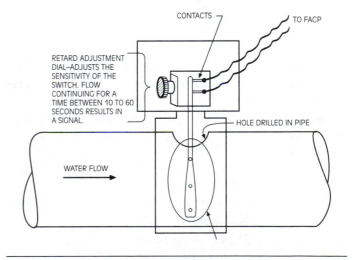

Figure W-3B   Vane type of water flow alarm.

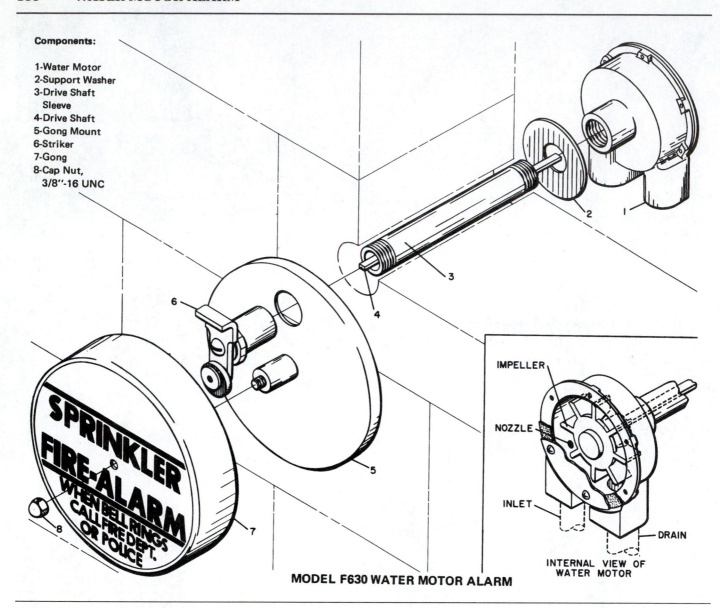

Components:

1-Water Motor
2-Support Washer
3-Drive Shaft Sleeve
4-Drive Shaft
5-Gong Mount
6-Striker
7-Gong
8-Cap Nut, 3/8"-16 UNC

MODEL F630 WATER MOTOR ALARM

INTERNAL VIEW OF WATER MOTOR

IMPELLER
NOZZLE
INLET
DRAIN

Figure W-4 Water motor alarm or gong.

Water Motor Alarm A mechanical bell alarm that is hydraulically operated by water pressure (Figure W-4). Also called a Gong.

Water Tender Any ground vehicle capable of transporting specified quantities of water. The National Wildfire Coordinating Group (NWCG) identifies three types (Figure W-5A). The Type 1 has the largest capabilities (Figure W-5B), the Type 2 (Figure W-5C) has less capabilities, and so on. Some departments call a water tender a tanker.

Water Thief See *Hose Appliance*

Water Vacuum See *Salvage Vacuum*

Web Gear A suspender-type webbing that firefighters use to hold fire shelters, day packs, canteens, and flares. Additional pouches can be added to carry radios, weather kits, flashlights, and so on (Figure W-6).

Webbing Nylon strapping, available in tubular and flat construction (Figure W-7).

Wedge Metal, wood, and plastic tools used for operations when material is to be spread apart (Figure W-8). Wedges are used when cutting trees and wood and during cribbing operations at a rescue scene.

Wet Barrel Hydrant See *Hydrant*

Wet Bulb Thermometer In a psychrometer, it is the thermometer with its bulb covered in a jacket of clean muslin that is saturated with distilled water before an observation (Figure W-9).

Components	Water Tender types		
	1	2	3
Tank Capacity (Gallons)	5,000	2,500+	1,000+
Pump Capacity (GPM)	300+	200+	200+
Off-Load Capacity (GPM)	300+	200+	200+
Max. Refill time (Minutes)	30	20	15

Figure W-5A National Wildfire Coordinating Group (NWCG) requirements for water tenders.

Figure W-5B Water tender—Type 1.

Figure W-6 Web gear with day pack attached.

Figure W-5C Water tender—Type 2.

Figure W-7 Webbing.

W

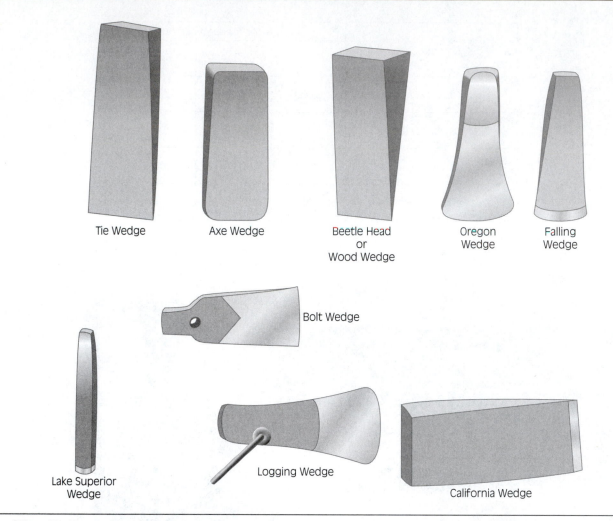

Tie Wedge

Axe Wedge

Beetle Head
or
Wood Wedge

Oregon
Wedge

Falling
Wedge

Bolt Wedge

Lake Superior
Wedge

Logging Wedge

California Wedge

Figure W-8 Various types of wedges.

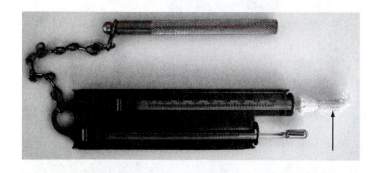

Figure W-9 Wet bulb thermometer.

Figure W-10 Winch.

Winch A pulling tool consisting of a length of steel cable around a motor or hand-driven drum. These are most commonly attached to the front or rear of a vehicle. Some may also be hand-cranked (Figure W-10).

Woven Jacket Hose See *Hose*

Wrecking Bar See *Bar*

Wye See *Hose Appliance*

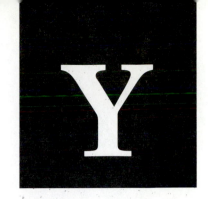

Yates Truckee Belt An emergency escape harness and tool belt that can be converted into a Class II harness (Figure Y-1).

Yes Tool A non-sparking, hatchet-type tool that is used for hazmat, vehicle rescue, and firefighting. It is a combination tool with a crash ax, hydrant wrench, spanner wrench, pry bar, hand pick, and gas shut-off. A yes tool is made of beryllium copper (Figure Y-2).

Figure Y-1 Yates truckee belt.

Figure Y-2 Yes tool.

Z Drag   A pulley system that is made from the same rope as the main line (Figure Z-1).

Zak Entry Tool   A forcible-entry tool with a hammerhead design, made of aircraft stainless steel. This tool is designed to bend but not snap if excessive force is applied (Figure Z-2).

Figure Z-2   Zak entry tool.

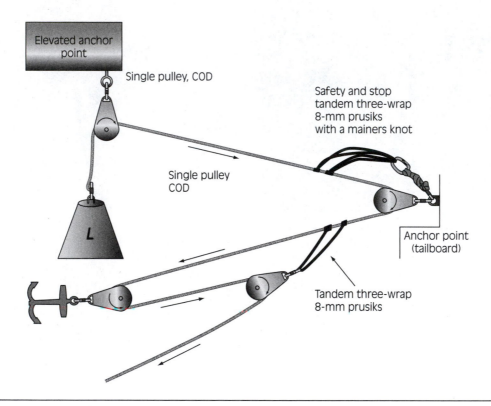

Figure Z-1   Z-drag, illustrating the 3:1 pulley system.